Fulgurites

Their Science, History, and Discovery

By Matthew Pasek, Ph.D.

Fulgurites by Matthew Pasek
Published by Free Radical Consulting
2703 Pemberton Creek Drive, Seffner FL 33584

© 2017 Matthew Pasek

All rights reserved. No portion of this book may be reproduced in any form without permission from the publisher, except as permitted by U.S. copyright law. For permissions contact:

freeradicalconsulting@gmail.com

Contents

1. Introduction 4
2. What is a fulgurite? 5
3. What is the history of fulgurites? 8
4. What other natural glasses are there? 17
5. How do fulgurites form? 25
6. What is their value? Where can I sell them? 30
7. Where can I find a fulgurite? 31
8. Where to find fulgurites part 2. 33
9. What objects look like fulgurites but aren't? 40
10. Fulgurites in Pop Culture 44
11. Fulgurites: the Science 45
12. What types of fulgurites are there? 48
13. Weird Fulgurite Minerals and the Rule of Threes 54
14. The special case of phosphorus 58
15. Reduction in Fulgurites 60
16. Using Fulgurites to Calculate Lightning Energy 64
17. Fulgurite Glass Density 68
18. Fulgurites are Shocking 69
19. How old can fulgurites be? 73
20. What are some of the minerals in fulgurites? 78
21. What type of rock is a fulgurite? 83
22. What is the difference between natural and artificial fulgurites? When might that be important? 84
23. What future science might there be for fulgurites? 88
24. Non-scientific uses 91
25. Individual fulgurite discussions 92

1. Introduction

I've written this book to fill a void on the rock enthusiast's bookshelf. There are numerous books written on meteorites, on fossils, and on superhero science. I've worked as a scientist on fulgurites for about a decade now, and have found these objects to be quite beautiful, and they can certainly capture the attention of the public. Not quite as much as a meteorite might, but definitely more than the average granite. Additionally, for my own edification, I figured a review of these objects might be interesting.

This book is a product of love and time. I'm not really that much of a cartoonist, or a photographer, but I've added both my comics and my photographs to this work. In both cases they may not be well done, the jokes may fall flat or be darker than is appropriate, and the figures will certainly appear unpolished. That said, they should illustrate the points they are trying to make. They were fun to draw, and I hope you find them interesting, and helpful.

2. What is a fulgurite?

Most people first learn of fulgurites when they step into a well-stocked rock shop. In between the more obvious merchandise--the shark teeth, quartz crystals, and petrified wood--a lucky few may find a sign advertising "petrified lightning!" Maybe this is accompanied by a brief description, perhaps, "A fulgurite forms as lightning crystallizes the sand it strikes. Metaphysical aspects of fulgurites include the ability to regrow hair and cure the gout." Although the latter maybe suspect, the former certainly has its appeal. Holding a sample of lightning—one of nature's most awesome destructive forces—has a WOW factor to it. Maybe some of that energy is still in there and can serve as a pick up like a cup of coffee when you're tired if you wear it around your neck? Maybe there's diamonds inside? The possibilities are endless. Perhaps, with a dose of skepticism, a friend suggests that a geology textbook might have some of the answers. If you look through your college textbook, you'll probably be disappointed with what you find. For the most part, they are mentioned in passing if they are mentioned at all. This book is here to fill the void and provide a few answers to some of your questions about fulgurites, both the common questions asked, the uncommon questions that should be asked, and a few questions that no one will ask but that I'll provide the answer to anyway.

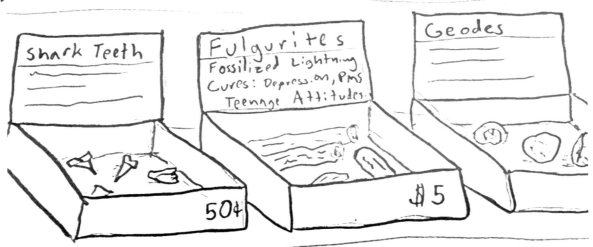

A fulgurite is, broadly speaking, a rock. A rock is a solid mixture of one or more minerals or mineraloids, typically inorganic, and which is part of the earth's crust. That's a mouthful, and some of the specifics of this will be given better detail later. In general, rocks are familiar to each of us. Rocks can be very large to very small. They can be the size of continents, down to the size of grains of sand. Rocks are separated according to their chemical composition, the minerals found within them, and the size and shape of individual mineral grains within a rock. The study of rocks is known as *petrology*.

A fulgurite is a special type of rock, called a "metamorphic" rock. Geologists, who study the earth, group rocks into three main categories: igneous rocks, which come from molten rock, sedimentary rocks, which are particles of rock that are glued together called sediments or which are minerals precipitated from the sea, and metamorphic rocks, the rocks formed from heating and/or putting other rocks under pressure, a process that forms a new, changed rock replete with its own minerals, textures, and crystals.

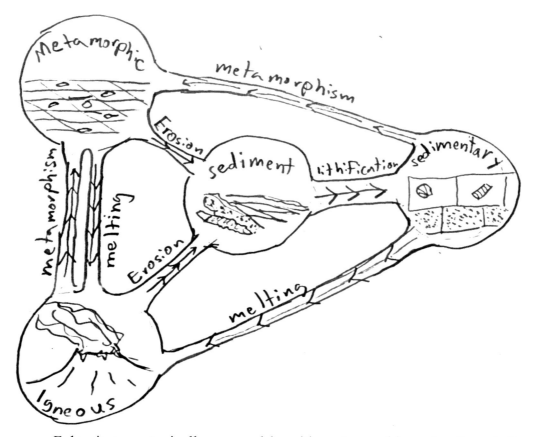

Fulgurites are typically grouped in with metamorphic rocks, though they don't quite fit this designation, as we will see later in this book. What really separates a fulgurite from other types of rocks is the presence of glass. Glass is defined as a semi-solid material (sometimes called semi-liquid) that does not have a repeating chemical structure. Glass is not a crystal. A crystal has a repeating, ordered chemical structure (generally... note that the exceptions to the rules often make for fascinating stories), and hence a fulgurite by definition is not a crystal. There are relatively few glasses that are found in nature, since crystalline structures are generally a more stable form.

Even more specifically, a fulgurite is formed by an electric discharge that melts rock to make glass. An electric discharge you are probably familiar with is lightning. However, fulgurites are also formed intentionally and unintentionally by humans. Intentionally formed fulgurites include those formed by scientists and science-enthusiasts, typically by operating high-power electrical equipment coupled to a sense of flagrant disregard for bodily harm[1]. Fulgurites may also be formed by shooting model rockets into thunderclouds, in an apparent effort to launch a war with the cumulonimbus cloud people. Unintentionally-formed fulgurites occur when power lines or other electrical equipment discharge electricity into soil or sand by an arcing current. Sometimes such occurrences can be found with the charred remains of the squirrel culprit who initiated the arc.

---

[1] See for instance, the formation of fulgurites by the Geek Group, who operate out of Michigan: https://www.youtube.com/watch?v=SMlW1APWTVk

This definition of fulgurites then does not require fulgurites to come from lightning. Fulgurites can be divided into natural and artificial fulgurites, and natural fulgurites do appear to require lightning for their formation. However, artificial fulgurites may form by arc welders, discharges from large capacitors, or downed power lines, in addition to lightning.

In general, fulgurites are uncommon. They can be quite hard to find. However, they are only uncommon because they tend to be small and fragile. Fulgurites should occur wherever lightning occurs over ground, and lightning occurs over almost all continental land masses, from Africa to the Americas to Asia to Australia. Antarctica may be the only exception, since lightning tends to require storm clouds to form, and cloud-to-ground lightning doesn't travel well through snow. However, Antarctica has a very small population of scientists and unlucky military personnel and most of these people are localized in one main region (McMurdo Station), hence not many people are free to observe an occurrence of cloud-to-ground lightning elsewhere on the continent. Beyond Antarctica, we can safely say most inhabited areas will experience lightning at some point. There is a reason that lightning safety is fairly universal across cultures.

The name fulgurite comes from the Latin word for lightning, *fulgur*. Although an alternative spelling, *fulger*, means the same thing in "vulgar" Latin, scientists and mineral collectors have by and large settled on the -ur spelling. A fulgarite is certainly a misspelling, and thus far has only been used by young earth creationist publications, perhaps tellingly so. My uncle has called these things, in a fit of genius, "boogerites," which is sadly incorrect.

3. What is the history of fulgurites?

Fulgurites have likely been known since antiquity. Lightning has long been associated with incredible things. Its power, the arbitrary nature with which it struck down people, and the explosive thunder that co-occurred with it gave lightning a supernatural character. Within the Judeo-Christian tradition, there is an association of divinity with lightning. Several of the most powerful gods in other ancient pantheons yielded lightning specifically as a weapon. From Zeus to Thor to Raijin, lightning was seen as the choice of the gods to smite the foes of the gods, which often meant whoever happened to be on the tallest hill during an ill-advised jaunt during a thunderstorm. Justification would usually come later ("Leodamus? Oh yeah, Zeus smote him because he overcharged for his goat cheese. That sinner."). Anecdotally, there are claims that fulgurites were considered to be the actual lightning bolts themselves, sort of like bullets, cast down during storms. The first serious scientific inquiry never really pursued this idea, since it was fairly obvious that hot temperature could make the glass seen. That lightning set fire to things was testament to its hot temperature.

The association between lightning and fulgurites began to be concrete by the 18th century. Lightning became an important subject of study around this time as well, at the time lightning and thunderstorms were major destructive forces causing huge amounts of damage by setting buildings on fire. Benjamin Franklin's invention of the lightning rod would eventually fix that, to the point where we don't tend to think of lightning as being very damaging to property outside of knocking out power to our homes occasionally. Nonetheless, lightning was a subject of much scrutiny in the 18th and 19th centuries, as buildings from churches, which tended to be the tallest buildings in town, to storage sheds containing gunpowder and exploding half a city were ignited by its blast. Lightning was the topic of many discussions at scientific societies about the characteristics of lightning, and fulgurites ended up getting a few mentions at the time, too.

One letter, sent in 1759 by William Mountaine to the Royal Society of the United Kingdom[2], described how a group of houses on Goat Street were struck by lightning during a storm. Mr. Mountaine describes visiting three of these houses, and surveying the damage. He observed that the lightning had shattered fine china and windows, propelling shards of glass from the latter outward with such force they had been lodged within the wooden door. The lightning had apparently traveled through some copper bell wiring, then bounced off to strike a nearby captain, completely missing his son who was holding onto his clothes (height matters!). The captain was knocked senseless for a day or two. A fair bit of time in Mr. Mountaine's description was spent on describing the rupturing of several flasks of port wine, which were stored in the house across the street from the three struck. Mr. Mountaine and the owner of the wine sampled the wine that had apparently been struck by the lightning (or shattered by the thunder shockwave), finding that wine to be rather flat. They then sampled a flask that had avoided being shattered in the storm, finding it "in perfection". This would have been the "control" sample, allowing the observers to differentiate between lightning wine and normal wine. A third bottle of wine, also sampled, was spared the calamity at a neighbor's house. By that time, the note-taking stopped, having sampled a number of wine bottles to ensure their perfection. The last one probably couldn't be called a control, though. Sampling wine happens often in science, usually when successfully winning a decent grant or getting a paper accepted for publication. Liquor is saved for not winning a grant, or for getting reviews back on a paper.

---

[2] See Mountaine, W., & Knight, G. 1759. An account of some extraordinary effects of lightning, in a letter to Dr. Gowin Knight: by Mr. William Mountiane, F. R. S. *Phil. Trans.* 51, 286-299.

A second letter to the Royal Society in the same century, by a Dr. Withering, discusses a lightning event with considerably less revelry[3]. Dr. Withering observed a storm, which struck and set fire to a cornfield, only to have rain put it out. Lightning from the storm then struck an old oak tree. Unfortunately, an unnamed gentleman had taken shelter from the storm under the tree. He was struck dead instantly. Some spectators ran to assist him when they saw him fall, only to find half of his clothes had been instantly ignited and burned off of his body. The earl who owned the land decided to erect a monument at that location to warn people of the dangers of taking shelter under trees during thunderstorms. Intriguingly, when digging the soil to erect the stone monument, a fulgurite was discovered. The top of the fulgurite corresponded to where the unfortunately gentleman had been holding on to his walking stick. This was the first scientific description of a known fulgurite-forming strike, as it was pretty clear that the lightning and formation of a fulgurite were closely related.

---

[3] The original account can be found searching for: Withering, W. 1790. An account of some extraordinary effects of lightning. By William Withering, M. D. F. R. S., *Phil. Trans. Roy. Soc. Lond.* 80, 293-295.

Lightning strikes trees, leaving scars.

The name "fulgurite" was first provided by Dr. Dominique Arago in an early work in 1821. Dr. Arago researched storms, thunder, and lightning, and fulgurites were a natural part of his oeuvre[4]. Arago's new name for these materials replaced their prior name of "fulminary tube" that had been used prior to this work. Arago and his contemporaries describe the first discovery of a fulminary tube as having occurred in 1711, by a clergyman named David Hermann. Pastor Hermann, finding these objects in the sand, did not really know what to make of them. His

---

[4] It's kind of hard to track down, but you can find the article in the original French as: Arago, M. 1821 Sur des tube vitreux qui paraissent produits par des coups de foudre. Annales de Chimie et de Physique 19, 290-303.

initial account suggested they were soft at depth but hardened on excavation. He also believed them to be an excellent remedy for fevers.

Recognition of these objects as forming by lightning was first made by a German, Dr. Hentzen. Additionally, with the invention of batteries in 18th century, and their improvement in the 19th, the first fulgurites were produced artificially by storing and discharging huge amounts of electric charge into sand. This conclusively linked fulgurites to electric discharges once and for all[5].

Probably the most famous scientist to investigate fulgurites was Charles Darwin[6]. Darwin, on the his voyage on the HMS Beagle, surveyed South America en route to the Galapagos (where he came up with a much more famous idea), and described the first fulgurites outside of Europe, in Maldonado in what is now Uruguay. There were several fulgurites in the region Darwin studied, and Darwin attributed these to lightning strikes that split in two prior to penetrating the ground. Additionally, some of these fulgurites bore "wings" which Darwin attributed to the plasticity of sand, suggesting that after the lightning strike formed the fulgurites, the ground shifted and squished a portion of the fulgurite, making them a bit more pancake-like.

A fulgurite likely similar to those seen by Darwin (from Texas, in this case).

After Darwin, fulgurites appear frequently in the scientific literature. For the most part, papers written about these objects identified them as coming from lightning as this was now

---

[5] The original articles are hard to find, but a review is given in a book by Sestier, F., 1866. La foudre. De ses forms et de ses effets.

[6] Darwin's account can be found as: Darwin, C., 1833. Fulgurites from Maldonado, South America. Voyage of H.M.S. Beagle. 53-54.

well-established, and most descriptions consisted of "Hey, we found a fulgurite here!" Fulgurites were found on most every continent, and a few were analyzed in depth. For the most part, however, the new science coming out of these objects was rather limited. It was clear that they were made of glass, and that compositionally they were almost the same as the material they formed from. Thus, the literature focused on a few things: where they were found, what their shape was, and differences in form. Two types of fulgurites were identified: sand fulgurites and rock fulgurites. Sand fulgurites formed thin glassy tubes, whereas rock fulgurites formed on mountain precipices, either as glassy glazes, or as "streamers" of glass through the rock.

Thus from the 19th century and for the majority of the 20th century, fulgurites in scientific study were quite boring. Fulgurites were made of glass. Glass is also found forming from volcanoes, and that glass is a whole lot more abundant. Sure, fulgurites were made of a slightly different glass than volcanic glass, but not different enough to make them really interesting.

The most interesting work at the time came from the new study of planetary impact craters. Impact cratering events were demonstrated to form glass, and several scientists included fulgurites in their analyses of natural glasses along with impact glasses. These scientists demonstrated that fulgurites were somewhat more closely related to impact glasses than to volcanic glasses. Mainly, fulgurites had a broader range of compositions. Volcanic glass came only from volcanic rock. Impact glasses also had a broad range of compositions, as impacts happen on all sorts of terrains[7].

Starting in the 1980s, some more detailed analyses of fulgurites were launched. The most significant new fulgurite science came from Eric Essene and Daniel Fisher from the University of Michigan, who were lucky to procure a fulgurite from Winans Lake, north of Ann Arbor. This fulgurite happened to form on a *moraine*, which is a geologic deposit formed at the edge of a glacier, and which is typically a mishmash of rocks that were crushed together as the glacier carved through it. The moraine was deposited in the ice age, but in the early 1980s, lightning struck along a tree at Winans Lake, forming a fulgurite. Upon closer investigation, Drs. Essene and Fisher discovered that this fulgurite had some of the most unusual mineralogy ever seen in a natural earth rock[8].

The fulgurite was large, having formed over had formed along a tree root, and was about 30 yards (~meters) in length, although present only intermittently over this distance. The most unusual feature, other than its size, was the presence of fairly large metallic grains within the fulgurite. These metals were composed of both iron metal and the element silicon, making minerals that were mixtures of the two.

Naturally occurring metals are fairly limited on the surface of the earth. There are relatively few: gold, silver, and copper are three many people might be able to name of the top of their head. Platinum and nickel can also form native metals, or metal alloys (mixtures of metals), under certain rare conditions on the earth. Notably absent from the list of naturally occurring metals is iron, which, outside of a few nickel alloys, does not occur very frequently at all on the surface of the earth. This may be surprising given how abundant iron metal is in

---

[7] See for instance, Bouška, V., & Feldman, V. I. (1994). Terrestrial and Lunar, Volcanic and Impact Glasses, Tektites, and Fulgurites. In *Advanced Mineralogy* (pp. 258-265). Springer Berlin Heidelberg.
[8] The original paper is Essene, E. J., & Fisher, D. C. (1986). Lightning strike fusion: extreme reduction and metal-silicate liquid immiscibility. *Science, 234*(4773), 189-193.

industry. We use it for building and construction, and for cars and machinery. Why then is it so rare?

Iron is indeed an abundant metal. It is the most abundant *transition* metal, those metallic elements that make up the middle plateau on the periodic table of the elements. However, most iron on the earth is tightly bound with the element oxygen. This chemical bond forms something less desirable: rust. This is the inevitable fate of all iron metal on the surface of the earth: rusting to make iron oxides, which may sometimes mix with water to make iron hydroxides. The process of rusting releases energy to the environment. Hence iron oxides are more stable than iron metal, and thus iron metal is rare.

Elemental silicon is even more problematic. Silicon should be familiar as the element that gives California's economic powerhouse valley its name. Yet elemental silicon is even more inseparable from oxygen. Any silicon you find on the earth's surface is likely bound to oxygen to make the compound SiO_2, the most common variety of which is the mineral quartz. Quartz makes up the sand on beaches, and is one of the most common minerals known to most people. It's really hard to pull off two oxygen atoms of quartz, and prior to the discovery of the Winans Lake fulgurite, the only place where silicon did not occur with oxygen was the depths of space, and even there, the separation between silicon and oxygen is less extensive than was found in the Winans Lake fulgurite.

Thus the discovery by Drs. Essene and Fisher provided for the first time a demonstration that fulgurites were unique. They weren't just analogs of obsidian and impact glasses, sandwiched compositionally between the two. They were weird.

The researchers suggested that the reason why the metallic iron and silicon were present in the Winans Lake fulgurite was due to the lightning strike following a tree root. Trees, or more specifically the wood that makes them, are made of the element carbon (and hydrogen and oxygen). When the lightning traveled along the tree root, it ignited the tree root and burned up the carbon. The carbon latched on to any spare oxygen that happened to be lying around (from iron oxide or SiO_2), and made these metals and native elements as a result.

This process is a natural parallel for how we humans make silicon and iron metal from ores. Ores of iron, such as the iron oxide minerals hematite (Fe_2O_3) and magnetite (Fe_3O_4), are burned with carbon, supplied as charcoal, and sand or limestone (the latter two to pull of contaminant elements), to make raw iron metal or silicon. This process is called ore smelting. The discovery of smelting was extremely important to dividing the human time line between the Stone Age, the Bronze Age, and the Iron Age.

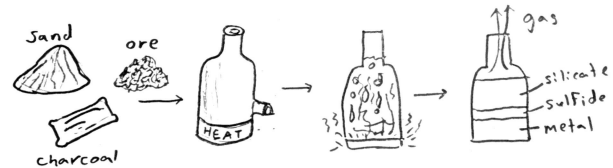

Rough diagram of how smelting works. Heat is supplied to the mixture of ore, sand/limestone, and carbon, which then segregate the elements in the ore into metals, sometimes sulfides (if the ore has sulfur), silicate that is termed slag, and gas, which is lost from the furnace. A pipe on the bottom of the furnace can be opened with liquid metal removed for further processing.

This discovery demonstrated that fulgurites were weird rocks. The Winans Lake fulgurite was especially strange, and since then few other fulgurites have shown the widespread formation of these metals. However, this work helped bring fulgurites back to spotlight as being rocks worthy of study. The next significant work on fulgurites came from Peter Buseck and Peter William's labs at Arizona State University. These researchers looked at the carbon molecule chemistry of a suite of 5 fulgurites, including the Winans Lake fulgurite. They discovered, in one of them, the presence of buckminsterfullerenes[9].

A buckminsterfullerene, a word more commonly shortened to fullerene or buckyball, is a molecule made of pure carbon. The elemental forms of carbon that are indubitably more familiar to you are graphite and diamond, both of which have been known since antiquity. In contrast, fullerenes have a much more recent discovery. The first discovered variety had a formula C_{60}, or 60 carbon atoms bound together to make a structure. The structure they made resembled a soccer ball, or a geodesic dome. Geodesic domes were a specialty of the architect Buckminster Fuller, and these molecules were shown to have a similar structure (think of Spaceship Earth, the central attraction at Epcot Center at Disney World). Carbon spontaneously assembles into these structures when heated rapidly and/or intensely. Fulgurites were among the first materials demonstrated to have these molecules. They were later also found in impact rocks and in forest fire soot.

---

[9] You can read the original scientific work here: Daly, T. K., Buseck, P. R., Williams, P., & Lewis, C. F. (1993). Fullerenes from a fulgurite. Science, 259(5101), 1599-1601.

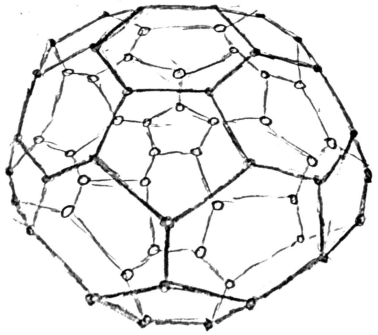

Schematic of a buckyball, which is actually quite hard to draw well.

This history brings us up-to-date on fulgurites through the 20th century. They are many other discoveries dealing with fulgurites through this time frame, but these may be viewed as the most seminal of the time.

4. What other natural glasses are there?

A fulgurite is distinguished from other types of rocks by the presence of glass. There are relatively few rocks that have glass inside of them. A glass is a disordered or semi-ordered, super-cooled liquid. Almost by definition, a glass is unstable: the material making up a glass wants to be ordered, getting its atoms in line. Naturally-occurring glasses are hence relatively uncommon. This is because glasses must form by an increase in temperature that exceeds the melting point of the target material, followed by a rapid cool down, fast enough to prevent the atoms from rearranging back to an ordered structure. This process is known as *quenching*. Conditions that allow a glass to be quenched into its disordered state are pretty uncommon as, if you encounter a high temperature environment on the earth, that environment usually persists for some amount of time instead of cooling off quickly. There are a few processes that have this feature, however. These are volcanic eruptions, rapid rock slides, and meteorite impacts, and, of course, lightning.

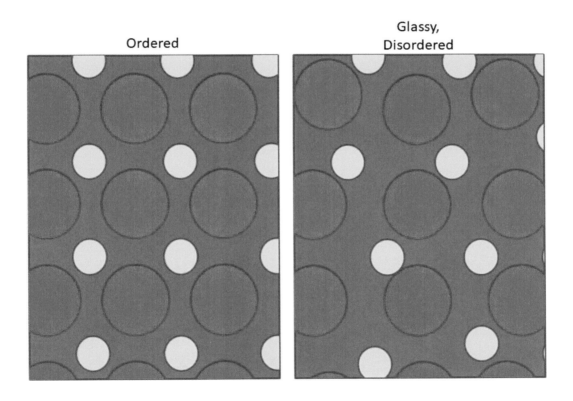

Schematic drawing of how a crystalline structure and a glass may differ.

Volcanic glass is probably the best known of all natural glasses. It is better known by the name *obsidian*. A stuffier name for this material is *tachylite*, and this name is usually confined to academic articles. Obsidian forms when volcanic rock cools so quickly that individual crystals within the rock don't have a chance to form. Obsidian was valuable to ancient cultures because when it was chipped just right, shards of obsidian have extremely sharp edges capable of cutting through many things. In fact, some modern surgeons have taken to using obsidian blades over steel blades as the edge is significantly sharper, and hence better able to cut through flesh.

Obsidian, from nearby Newberry Volcano in Oregon, USA.

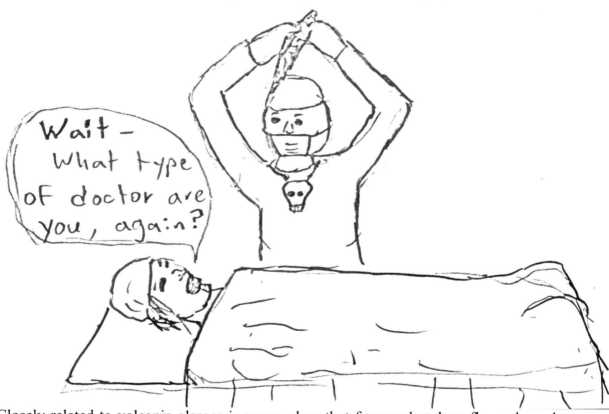

Closely related to volcanic glasses is a rare glass that forms when lava flows through a rock and ignites carbon residues within the rock, heating the rock past its melting point and forming glass, a variety of which is called *buchite*. Such an occurrence is not common, but forms one of the few types of natural glasses.

Pseudotachylite is not a common rock in most rock shops, but is another naturally occurring glass. This glass, which is typically dark and obsidian-y in texture, is not volcanic in origin. Hence the name, false-obsidian, or pseudotachylite. This glass is now believed to form when a rock slides against another rock really fast. If you picture what happens to your leg when you slide against the ground trying to reach first base in baseball or softball while wearing shorts, you have some idea about the formation of pseudotachylite. When a big block of rock slides against another block of rock, it heats up. If the rock stops sliding suddenly, such as at the end of an earthquake, it can cool quickly and form a glass.

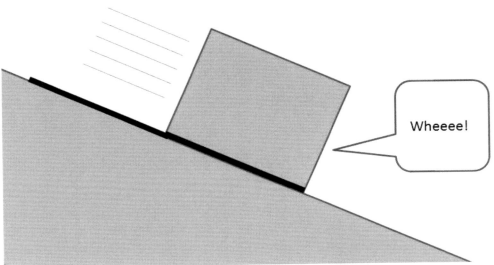

Schematic of pseudotachylite formation. Note that rocks don't "experience" metamorphism; they enjoy metamorphism.

Much more common in rock shops—and often misidentified as meteorites—are the impact glasses. These glasses formed when a giant meteor crashed into the earth, creating a fireball that contains a mix of the target rock and of the meteor, though mostly the former. In contrast, a meteorite is an actual rock from space, and not just melted earth rock. Nonetheless, these glasses are quite unusual and indicate major extraterrestrial events. They range greatly in size, color, shape, and composition. The easiest to acquire are the tektites, which are black, knobby, sometimes drop-shaped, and often chipped to expose the glass underneath. Most of these come from Southeast Asia or Indochina. A few of the more rare types come from Australia, where unique splash-form buttons can be found. Tektites are fairly young impact glasses, as they are less than a million years old.

Tektite from Indo-China. Width is about 4 cm.

Note that the term "tektite" is often used to describe many impact glasses that have been found, including several that are definitely not associated with the <1 million year old Southeast Asian impact. The crater associated with this impact has yet to be found, however, several other glasses have been found and are associated with impact craters. One type of impact glass that is fairly distinct from the black tektites is Libyan Desert glass. This glass is pale yellow in color, and is also the most transparent of all natural glasses. The surface of Libyan Desert glasses are often pitted, almost etched, much like beach glass. Libyan Desert Glasses are composed mostly of silica, with a formula of SiO_2. Samples of this glass have even been found as part of the jewelry of Pharoahs. These glasses have been estimated to be about 30 million years old.

Libyan Desert glass. Length is about 2.5 cm.

Some of the prettiest of impact glasses are the moldavites, which are mostly found in the Czech Republic and in Germany. These glasses, which are bottle-green in color, tend to be more jagged than the common Indochina tektites. They can be faceted like gemstones. These glasses are about 14.5 million years old, and are associated with the Ries impact crater in Germany. In addition to the moldavites, there are a half-dozen or so other impact glasses that likely formed when a large meteor struck the earth, vaporized the ground in a giant fireball, and then rained out glass as that vapor cooled and solidified. These impact glasses tend to be aerodynamic in color, and often don't look like the rocks or soil they are found in.

A translucent moldavite. Length is about 2 cm.

Another variety of impact glass falls better within the definition of pseudotachylite. These glasses are associated with craters. Unlike the tektites (including moldavites and others), these are not shaped aerodynamically, but form large masses of glass. These were likely formed as rock was propelled against itself in an impact, liquefying but not vaporizing it. These glasses are much less commonly sold in rock shops (because they're not really that pretty), but still illustrate the incredible power of an impact crater-forming event.

An impact glass from Zhamanshin crater, in Russia. The width is about 12 inches. The lines are due to flow of the glass after the rock melted. This rock was sliced from a much larger sample, and cut to a thickness of about 0.75 cm (about 0.3 inches).

Sometimes glasses are not formed naturally, but are formed unintentionally by man. One of the most unusual human-produced glasses was formed in large masses in 1945, near Alamogordo, New Mexico. If you remember your history, this was near the end of World War Two. New Mexico was the testing site for the first atomic bomb. An intriguing side product came from mankind's first demonstration that it was now capable of wiping itself off the face of the earth: a light gray-green glass called trinitite. The atomic fireball vaporized a few acres of New Mexico when the bomb was dropped. Naturally, rock doesn't make a good gas, and hence the gas cooled, dropping molten bits of glass from the fireball down to the ground.

Of course, this being the 1940s, enterprising collectors grabbed a fair bit of this glass, and within a year, jewelry featuring the glass came out. Women wore this jewelry both as a celebration of both the US's triumph in World War Two, and because they enjoyed radiation burns, apparently. The US government later bulldozed the site to prevent these shenanigans, but not before a fair bit of material had been collected.

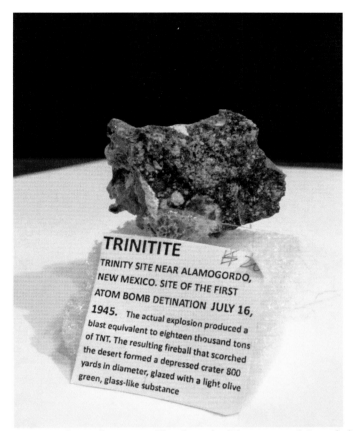

A piece of trinitite acquired from a gift shop in Tucson, with original mount and label (and price).

As you can tell, a connecting theme between the natural glasses (and trinitite) and fulgurites is that they were formed by high-power events. The lowest power one—volcanoes—is still pretty terrifying when you're close to it. Indeed, impact craters were first scientifically studied by investigating nuclear bomb explosions, which turned out to be a close analog. Within this crowd, lightning appears to be a much less terrifying thing to study!

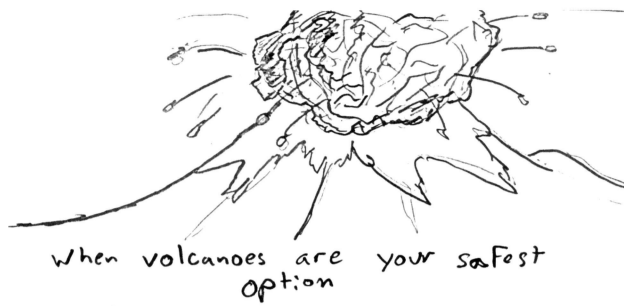

when volcanoes are your safest option

Fulgurites are glasses produced by lightning. In order to understand their formation, we must discuss lightning first.

5. How do fulgurites form?

Lightning can be described as an electrical *spark*. Electrical sparks are important to modern technology. A spark provides a small burst of energy that starts many modern conveniences, from your car's internal combustion engine, to lighting your low-power fluorescent bulbs. A spark is a non-continuous movement of electrons from one region or pole to another. This is in contrast to an electric *arc*, which is continuous. What is "continuous"? In this case of electricity, if you can keep the spark flowing for as long as you'd like, it's an arc. Naturally, "continuous" and "non-continuous" are in the eyes of the observer, and no single definition works that great for these terms. Lightning is also commonly called an arc, but it is pretty short in duration.

Sparks generally occur with insulators. A brief note on definitions: a *conductor* is a substance that allows electricity to flow through it, and an *insulator* does not. These definitions depend on a number of things (amount of current and voltage, temperature, level of acceptable heating, etc.), and like many things in science, there's not necessarily a clear cut break between a conductor and an insulator. Under some conditions, an insulator can act as a conductor. A spark generally occurs when electrons fly across or through a non-conducting material or insulator. This occurs when the insulating capabilities of the material "break down". In other words, things that don't want to let electrons bounce around on them will eventually do so when there's too much energy available to hold the electrons back any more. As long as electrons keep flowing through the substance, then you have an arc.

Sand, soil, and most rocks tend to be pretty good insulators. When electrons flow through an insulator, it necessarily creates a glowing spark or arc. A spark or arc glows because the electricity it is actually traveling through super-hot *plasma* that crosses the insulator.

When we learn about science in elementary or middle school, we are taught that there are three phases of matter: solid, liquid, gas. These are the three you encounter regularly in your everyday life: from the chair you are sitting on (solid) to the air you're breathing (gas), to the coffee (liquid) you're sipping (possibly). The forgotten fourth member that is now quite common in our lives today is plasma. Plasma is similar in many ways to gas, but where the electrons within gas are bound to atoms, in plasma, some number—sometimes all—of the electrons are not bound to atoms. They flow around freely, carrying electrical charge around easily. Plasma is a gas that hence conducts electricity, and behaves in many ways like a metal. Plasmas are found in fluorescent bulbs, spark plugs, plasma screen televisions, and lightning, so you'll note that the plasma as a familiar state of matter is more recent than solids, liquids, and gases. Also note that physicists and chemists have since identified several more phases of matter, but almost all of these exist only in weird cases, some of which are found only in laboratories.

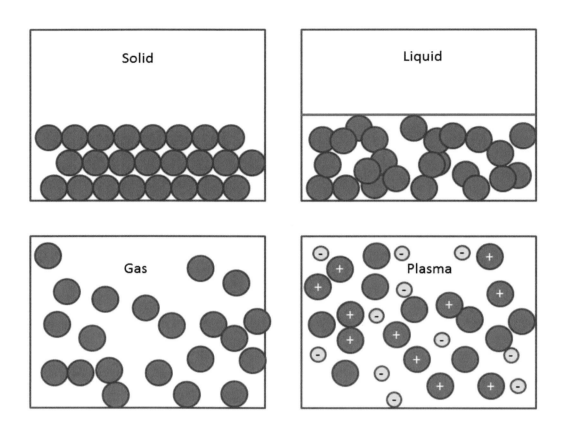

Plasma forms when there is sufficient voltage to force electrons to flow through an insulator to reach a conductor. The amount of voltage necessary depends on the thickness of the insulator (thicker insulators require more voltage) and the composition of the insulator.

A volt is something familiar to most of us. It's placed on the side of the batteries we buy, along with the size, for instance "AAA" or "AA". A volt is a unit that describes the amount of energy each packet of electrons delivers as they flow from one side of an object to another. In the case of a battery, this is the negative side of the battery to the positive side. A single volt is a joule, which is a unit of energy, per coulomb, which is a quantity of electrons. Energy is measured in joules, which itself is a unit that consists of mass times velocity squared (kg m^2/s^2). A joule is not a lot of energy; it is about the energy it takes a single drop of water to heat up by one degree. A four pound ball moving two miles per hour also has about a joule in kinetic energy. Getting hit with that ball wouldn't really hurt too badly.

Other measures of energy are also familiar to you. Calories are probably the most familiar of these units of energy. These are found on the back of your candy bar, or box of cereal. A Calorie is different than a calorie; the former is 1000 times the latter. A calorie is defined as the amount of energy it takes to change the temperature of one milliliter of water by one degree Celsius, and is about 4.18 joules. A candy bar may provide 300 Calories when you eat it. 300 Calories is about 1.25 million joules. We humans do require a lot of energy to do the things we do, especially with a 2000-3000 Calorie diet.

Another unit of energy is the kilowatt-hour, a unit that measures how much the power company is going to charge you on your next bill. A kilowatt-hour is 3.6 million joules, and you typically pay 12 to 20 cents for it. To reiterate: a joule is not a lot of energy.

The second unit that defines a volt is the coulomb. A coulomb is a simpler unit, as it's merely a set number of electrons. That set number is pretty big, though. Without using scientific notation, a coulomb is rather unwieldy. It is 6,242,000,000,000,000,000 electrons. In scientific notation, this is 6.242×10^{18} electrons, a bit more manageable.

So, one and a half joules is the energy every 6.242 quintillion electrons give off for going from the negative side of a 1.5V battery to the positive side. You can tell from this number that it takes a lot of electrons moving from one side of a battery to the other to make the energy needed to power things.

This number may seem huge, but let's put this in perspective. The average power line feeding into a house in the US has a voltage of 120 volts. The average house also uses about 1000 kilowatt-hours per month. Thus, to provide keep your house powered, the power company shuffles around 187,000,000,000,000,000,000,000,000 electrons (more easily read as 1.87×10^{26}) every month. The small size of electrons makes this number so very unnoticeable.

Lightning is a spark or an arc, depending on who you're talking to. In order for cloud-to-ground lightning to happen, there must be a difference in electric potential between a cloud and the ground. This difference comes from differences in charge between these two places. Something similar happens when you walk across a rug on a cold, dry night in socks. By walking, you knock off a few electrons from your socks and the rest of your body. When you touch something metallic later, you get shocked as the missing electrons are replaced by others that come bouncing back to cancel out this charge imbalance.

With lightning, this charge difference is much larger than what happens with you and your socks. In order to make electrons travel across a spark in the air, the voltage must be about one million volts for every meter or yard of distance traveled. Your batteries are about 1.5 volts, so the voltage for lightning, which must travel through the air often over lengths of miles, has to be a lot larger. This voltage is so high that some scientists have suggested that lightning is initiated when cosmic rays bombard the atmosphere of the earth. The lightning follows the trail of a cosmic ray, as the trail starts out somewhat ionized.

The hard part of making a lightning bolt is building up enough voltage to make the spark happen. Once a spark happens, there is a trail of plasma that provides the perfect stream along which electrical current can flow. This current typically flows from the negatively charged cloud to the positively charged ground, and this is called negative cloud-to-ground lightning (note: if lightning travels from a cloud to another spot in the cloud, or to another cloud, that is a lightning flash, not a strike). A lightning strike may consist of several strokes, which are multiple instances of electric current flowing from the cloud to the ground. If you've ever seen a lightning strike the ground, sometimes several strokes will occur over the course of a single second, perhaps with enough time between them to be registered by the brain as different instances. This may give the lightning strike a strobe-light effect.

A fulgurite is formed when lightning travels through the ground. It does so since beneath the ground there may be a conductive layer that allows for the equalization of charge. For many spots on the earth, this is a near-ground or subsurface aquifer (where groundwater is). When the lightning strike travels through the ground, it may follow the path of least resistance, such as a tree-root. But at the same time, the high voltage of lightning makes it similar to the joke about the eight hundred pound gorilla: where does it sleep? Anywhere it wants. Similarly, where does lightning go? Anywhere it wants. It may follow the line of least resistance, especially if that's substantially less resistive/more conductive than the material around it, but it will not necessarily do so.

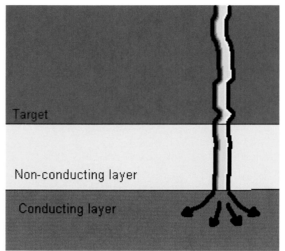

A really bad illustration of lightning traveling through soil to make a fulgurite.

When lightning travels through soil, energy is deposited into the ground. This occurs because of a process called electrical breakdown. A certain voltage is required to initiate a spark in soil, and this voltage is typically about one to ten million volts per meter or yard of soil traversed. As the first few electrons make this trek, they drop a lot of energy into the ground. The result of this energy is the rapid heating of the ground along the channel of the spark, which vaporizes a narrow tube within the ground, turning it into plasma. Electrons flow through this plasma, subsequently adding heat that warms the regions immediately adjacent to the tube.

After the lightning strike ceases, the hot plasma escapes from the tube quickly and the ground begins to cool down. It typically takes a few minutes cool to below the melting point of rock, and maybe up to twenty minutes to cool to a point where it may be picked up and held as a fulgurite. Because the rock cools so quickly, the melted portion of it will freeze into glass. The center of the rock is hollow, because that is where the lightning plasma flowed through it to reach the conducting layer, vaporizing a channel in the rock. Adjacent to this hollow void space is glass, a reminder that the rock was at one time molten. Glued to or embedded within the glass will be fragments of rocks or particles of the soil. These fragments may include sand grains for those fulgurites occurring within sand, or for those that occur in gravel, may be whole pebbles. Sometimes, some of the grains in this outer layer include material that partially melted, gluing the outer wall together as a crusty, gray material. We call this outer material *baked*, as this term describes how a good portion of this crust formed by drying out the rock and melting it along some edges of the grains. Note that a baked rock differs from a baked person, who may try to sell you on the metaphysical properties of a fulgurite.

A fulgurite from Greensboro, NC, showing a central void surrounded by gray-green glass and a baked exterior.

The shape of a fulgurite stores the history of the lightning that formed it. It is a rapid history, intense and powerful. It is one that tells the story of nature's power, a power beyond anything possible by humans until 70 years ago, when the first atomic bomb blew a hole in New Mexico. They hold a beauty, a beauty that comes with knowledge of what they mean.

6. What is their value? Where can I sell them?

A common question a non-geologist will ask a geologist is "How much is it worth?" Unfortunately, unless the object is an ore (coltan), a fuel (coal), or another fungible commodity (gold), the answer is an ambiguous, "Whatever someone will buy them for". Helpful, I know. But that's because the business of a geologist is not appraisal.

A fulgurite doesn't have a lot of intrinsic value. It is composed of glass, and though natural glasses are rare, they do not concentrate rare metals or nuclear elements any more than the soil they are found in. Glass is also quite common as a man-made substance. So, fulgurites don't really have a lot going for them. That said, in my quick eBay analysis, I've found an average price of about $1.50 per gram for the Type I fulgurites (details in chapter 12). These are the most commonly traded fulgurites. Types II and IV fulgurites are typically rarer with respect to availability, but conversely they are often cheaper since they are larger. At a $1.50/gram, there will be marked price variations, and the wholesale will typically be a lot lower for large lots. Selling a single fulgurite one at a time will make a lot more than an entire collection. Naturally, it will take a long time to liquidate a group of rocks that way.

Mineral-enthusiasts and rock collectors will often seek to procure at least one fulgurite for their collections. Beyond that, it takes someone who is at least a bit familiar with fulgurites, knows some of the distinguishing features, and is at least interested in what they are and how they came to be to really seek to acquire fulgurites (e.g., more than one). Someone like you, perhaps? Maybe after you've read this book?

Beyond a sense of familiarity, a collector of fulgurites may seek these things because they have an interest in acquiring rare terrestrial glasses. Often these people also like weird rocks like meteorites and tektites, and meteorite sellers often also have fulgurites for sale. Additionally, a well-characterized fulgurite may have a special appeal to a mineral collector. This is because fulgurites may have extremely rare minerals that aren't encountered in nature often. Some of these minerals include ferrisilide, hapkeite, schreibersite, xifengite, gupeite, and baddeleyite. They may only be visible in microscope, but to the completionist mineral collector, fulgurites are one of the rare places where some of these minerals may be found. That said, most fulgurites don't have these minerals, and it takes careful analysis to find one that does.

The largest market for fulgurites is the internet. Rock and mineral shows are also good places to trade fulgurites. In general eBay is great for getting a few simple pieces, though more unusual fulgurites are uncommon. Rock shops, when they have fulgurites, tend to fall into two groups. One type of rock shop often has unique pieces, typically more interesting than those you'd find online, often found by a local and sold for cold, hard cash. The second type of rock shop has bought a bunch from Algeria in bulk and is slowly liquidating them. Be forewarned that very few rock shops carry fulgurites regularly.

7. Where can I find a fulgurite?

If you're ready to try to find a kilogram (2 lbs) of fulgurites to pay the mortgage for a month, where should you go? Honestly, they can be anywhere. There might even be one in your backyard, waiting for you, right now…

A fulgurite occurs wherever lightning can strike soil. Lightning can occur almost anywhere, and soil is found almost everywhere. The exceptions to these include the oceans, which have no soil, lakes and rivers, whose soil is too far buried to be accessible, large cities, which have no soil, and the Arctic and Antarctic, which have little to no lightning.

With these no-fulgurite zones in mine, there's a chance to form fulgurites pretty much everywhere else. An exception may be forests. It's unclear whether fulgurites form in forested regions frequently, but given that there have been several reports of fulgurites forming along tree roots, it's pretty likely that there will be some, but they are probably buried.

The number of fulgurites that form each year can be calculated by multiplying a series of numbers together.

# of fulgurites = Lightning frequency × %CtG × Fulgurite forming fraction

First, the number of fulgurites that can form is proportional to how many lightning strikes hit a region. This number is fairly well known as an average by location. Check out the websites of Vaisala for a quick map, for instance. For most of the US, the southeast is the most lightning-prone. About 5 lightning strikes happen per square kilometer each year (about 10-15 per square mile) in the hot south. This is due to the thunderstorm-rich summers. The northeast is less prone to lightning, maybe 1 or 2 strikes per square kilometer. The Pacific coast and northwest have even fewer still, with between 0 and 1 per square kilometer on average. An exception here lies in the mountains of the west and northwest, where lightning will strike frequently and with loss of life (avoid mountain climbing if a storm is predicted!). The American southwest and Midwest have frequencies in between the relative calm of the Pacific coast and the thunderstorm-heavy southeast, at around 3 per square kilometer on average.[10]

In the USA, the state of Florida is the leader in lightning. Some counties have over 10 strikes per square kilometer. Florida does not lead the world, though. That honor goes to equatorial Africa as there are regions that can get over 50 strikes per square kilometer.

A fulgurite will form only through cloud-to-ground lightning. Cloud-to-ground lightning is about 25% of the lightning that occurs on the earth, and cloud-to-cloud or intra-cloud lightning is the remainder. This percentage does vary significantly with location- some places are less than 10%, some are 50%. The average does tends to be about 25-33%.[11]

The frequency of lightning multiplied by the fraction of cloud-to-ground strikes and then multiplied by the frequency that a lightning strike will form a fulgurite gives the number of fulgurites that form per year per area. This last value is harder to estimate. Experiments performed by University of Florida researchers showed that if lightning strikes a good target, such as sand, it will always produce a fulgurite. However, this number must consider the amount of trees and buildings in an area, since if they get struck the ground does not. Hence in cities and

---

[10] You can track the lightning that is occurring right now on: http://thunderstorm.vaisala.com/explorer.html
Alternatively, you can look at the yearly average at the bottom of the page here:
http://www.vaisala.com/en/weather/lightning/Pages/default.aspx

[11] Price and Rind discuss this in their paper: Price, C., & Rind, D. (1993). What determines the cloud-to-ground lightning fraction in thunderstorms? Geophysical Research Letters, 20(6), 463-466.

in forest, the amount of fulgurites that can form should be low to none. Thus, since the American Southeast is fairly tree-heavy, there are fewer fulgurites that form than would be expected. Contrast that with the American Southwest: with fewer trees, fulgurites should form with almost every cloud-to-ground strike.[12]

That's not to say that a fulgurite doesn't form when lightning strikes a tree. In fact, we know from history that fulgurites can be found at depth when lightning strikes a tree. Dr. Withering's story (see Chapter 2), for example, had a fulgurite that was about 18 inches below the surface. More recently, Eric Essene's study on the Winans Lake, Michigan fulgurite also formed when lightning struck a tree and traveled along its roots. However, it will probably be less obvious when a fulgurite is formed along a tree, because they might be deeper and harder to access than when lightning strikes soil. When the Winans Lake fulgurite formed, the ground exploded around it, so it may be reasonable that you might find a fulgurite by seeing what happens to the soil around trees after a lightning strike.

To summarize the number of accessible fulgurites formed for each area can be calculated from the formula above or maps available on the internet. Using the map, you can make a rough estimate. Take the number of lightning strikes per unit area where you're looking on the lightning map, then divide it by 4 for dry, treeless, uninhabited regions. If you're looking in a forest, take the number of lightning strikes per unit area and divide it by 12. In cities you're very unlikely to find a natural fulgurite, so dividing the number by 20 is reasonable for small cities (e.g., Tucson), or just assume zero for large cities (New York).

---

[12] The original paper is: Jones, B. E., Jones, K. S., Rambo, K. J., Rakov, V. A., Jerald, J., & Uman, M. A. (2005). Oxide reduction during triggered-lightning fulgurite formation. Journal of atmospheric and solar-terrestrial physics, 67(4), 423-428.

8. Where to find fulgurites part 2.

A major factor that determines where lightning strikes is the height of an object, hence knowledge of local topography is very useful for finding fulgurites. Many fulgurites have been found at the summits of mountains. This is because lightning likes to strike tall objects. You can visualize the rationale for this propensity by using a tool called the "Rolling Sphere" approximation. The "Rolling Sphere" method is used principally to protect against lightning strikes, but also can be used to predict where lightning might strike. What this rule effectively says is that lightning may strike at any spot where a large, imaginary sphere comes in contact with the ground. If you roll the sphere around in your mind, you can figure out safe spots during storms, or even likely fulgurite-producing zones.

To visualize this, imagine a sphere that is pretty big (about 40 yards in radius, double that for diameter). Roll it around, squishing nothing in sight, but coming in contact with the tops of everything nearby. The sphere will probably move around the edges of your house or other nearby buildings, will dance along the tops of trees, and will have its tallest point at the peaks of mountains. This tool is used to help assist in designing systems that provide lightning protection for sensitive electrical equipment, and has saved many millions to billions of dollars.

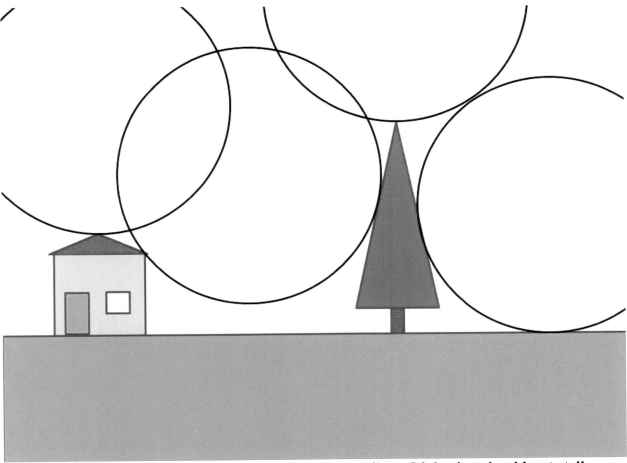

The rolling sphere predicts where lightning strikes. Lightning should not strike any region that the sphere can't contact, such as between the house and the tree (provided the object isn't tall enough to touch the sphere).

If the rolling sphere method isn't doing it for you, an alternative way of figuring out where lightning may strike is by using the 45° (degree) rule. Effectively, the tallest object viewed at 45° up from the horizon is the most likely to get struck. If there is no tall object, then you are the tallest object. In most cases, the tallest object will be nearby poles, trees, mountains, and buildings.

Using either the 45° rule or the rolling sphere model, we can infer that the top of a mountain is usually a good place to find fulgurites. A mountain is usually the tallest point in a region, and the tallest point is the most likely place for lightning to strike. That's not to say that lightning won't strike elsewhere- by the 45° rule, lightning will strike anywhere if the slope of the mountain is less than 45°. Most mounds and hills are about 30°, hence they may be struck anywhere along the slope. Additionally, another rule is that lightning does whatever it wants. Pictures of the Empire state building during storms show that lightning occasionally strikes it on the side, and not necessarily on the top. Most strikes still occur on the top, so these rules are sufficient for most cases.

No place is safe from lightning, especially when you're the tallest building in New York City.

The calculation given earlier gives you the number of fulgurites that form each year in a given location. However, fulgurites can accumulate over a number of years, and it's actually very difficult to determine if a fulgurite formed a few days ago or a few years ago. For this reason, the actual number of fulgurites that can be found in a location depends on three things: 1) the above calculation that gives the number of fulgurites per area per year, 2) the age of an exposed surface, and 3) the material the fulgurites are made of, which impacts their average "lifespan".

The second question then is: How long has the ground been exposed? For most natural settings, this is long enough to not matter. The times it does matter are at new construction sites, at beaches that have had their sand replaced to fight erosion, or fresh rock surface/roadcuts. In general, these places are too young to have accumulated any fulgurites. Hence if your house was built recently as part of a new subdivision, and the ground got torn up during its construction, it's probable that there are no nearby fulgurites. How old is old enough? That's an open question, but probably on the order of about 100 years will suffice for most places.

Given that most ground is suitably "old", the more important question is: What type of ground is it? The type of ground determines the type of fulgurite that forms there, and the average "life" of a fulgurite is determined by its type. Fulgurites formed from clay turn back into clay relatively quickly. Fulgurites formed from sand may take a while to covert back into sand. Fulgurites formed on rocks may also take a while to convert back into rock, depending on how much rain they get exposed to. The life of a fulgurite formed from clay may be only 10 years or less, which is pretty young if you figure the average ages of rocks are several millions of years. Fulgurites formed from clay have elements inside of them that want to reorder into non-glassy structures, and hence they fall apart pretty quickly.

Putting this all together, I've made a handy chart for figuring out how many fulgurites may be found on each square mile of land. This doesn't mean you'll find any- these things can be quite small and hard to spot. Still, it should give you an idea if it's worth looking for these things at all. Of course, the easiest way to procure a fulgurite is to see one form nearby, hopefully suitable secured inside a house or building.

Lightning Strikes per year	Forested?	Soil Type	Number of fulgurites / mi^2
0.1	N	Sand	600
0.1	N	Dirt	0.6
0.1	Y	Sand	200
0.1	Y	Dirt	0.2
0.5	N	Sand	3000
0.5	N	Dirt	3
0.5	Y	Sand	1000
0.5	Y	Dirt	1
1	N	Sand	6000
1	N	Dirt	6
1	Y	Sand	2000
1	Y	Dirt	2
2	N	Sand	10000
2	N	Dirt	10
2	Y	Sand	4000
2	Y	Dirt	4
4	N	Sand	25000
4	N	Dirt	25
4	Y	Sand	8500
4	Y	Dirt	8
6	N	Sand	40000
6	N	Dirt	40
6	Y	Sand	10000
6	Y	Dirt	10
8	N	Sand	50000
8	N	Dirt	50
8	Y	Sand	20000
8	Y	Dirt	20
10	N	Sand	60000
10	N	Dirt	60
10	Y	Sand	20000
10	Y	Dirt	20

Again, you can use the lightning maps from reference 10, which translate to the above in square miles. Note that if you'd consider the terrain more "rocky" take the dirt number and multiply by 10.

A second handy bit of information for finding fulgurites is knowing where to look. Generally, there are three ways to look around for fulgurites. The first, and best way, is to go where you've seen lightning strike. This will usually occur just after a storm is over, and be within viewing distance. Don't go out during a storm, for obvious reasons. Lightning can vaporize rock. Imagine what it can do to you.

Watching for lightning is obviously going to be a lot of waiting with a bit of luck thrown in. Additionally, if there are trees nearby, there's a good chance the lightning will strike those, and any fulgurites, if they form, may be deep near the roots of the trees (again, remember the case presented by Dr. Withering in the 18th century). That said, sometimes fulgurites will form at the surface of tree roots, and if you see a tree get struck by lightning, it may be worthwhile walking the perimeter to see if any fulgurites have formed. Search for dead grass a day or two later, as that may be a sign that the ground got baked and a fulgurite is nearby.

Seeing lightning strike ground nearby is going to be the easiest way of finding a fulgurite for most people. However, a second way of finding fulgurites is by going where they have a chance to accumulate. Two places where they will often accumulate are on mountain tops, and on sand dunes.

Mountain peaks are excellent places to find fulgurites because they are high points made of rock. Rock that turns to glass, when the glass is not extensively exposed to water, is quite stable, and hence fulgurites formed on mountains (which by necessity are dry, since water flows downhill) will collect and be pretty obvious. The most difficult part with collecting mountain fulgurites is their removal. Since they are often surficial melts, unless the rock has shattered, the fulgurite cannot be easily removed.

Another great place to find fulgurites is dunes. Sand dunes are typically topographic highs, which means they'll often get struck by lightning, provided storms are frequent. For dunes near beaches or the ocean, this should happen frequently. The height of the dune is important, as small dunes don't have the height or volume necessary to get hit much.

Equivalently, sand mines are often great places to find fulgurites. Sand mines provide sand for use in construction, golf courses, playground sand, and sometimes to even fill in beaches. Many sand mines are in fact ancient dunes. Since they were at one time dunes, they are often heavy with fulgurites. If the sand mine is active, sometimes the best place to find fulgurites will be in spoil piles. These piles are the left-overs from sifting the sand to purify it for shipping it to other places where it's needed.

A sand mine located in Polk County, Florida USA. Several Type I fulgurites were found here.

Notably absent from this list are the number one places people think about when they think about fulgurites: the beach. The beach is not a good place to find fulgurites, outside of the few times they form by an observed strike. This is because beaches typically don't have a lot of sand (compared to dunes), and most beaches are actually young. They are young because the modern day modification of the coast has resulted in stripping of beaches, hence many beaches are actually artificial, and have been recently replenished with sand from elsewhere, such as from sand mines. Additionally, beaches have waves and tides, and these are pretty aggressive towards fragile glassy fulgurites. Beaches shouldn't be at the top of the places you go to find fulgurites. You'd be better off searching in the desert, since the desert tends to be much less active, and, at least in the southwestern US, the deserts receive much of their rain from torrential summer storms, increasing the opportunities to form fulgurites.

9. What objects look like fulgurites but aren't?

Having made several purchases of fulgurites over the years, I've made some bad ones that emphasize "let the buyer beware". Several objects I've bought have turned out not to be fulgurites at all. They looked like fulgurites, at least as far as an initial purchase is concerned. On greater inspection these objects have turned out to be other geologic materials. In this section I'll provide you with information allowing you, the discerning reader, to be better equipped to purchase a true-blue fulgurite if the opportunity should present itself.

First of all, a caveat. A fulgurite is formed from a spark discharge traveling through soil or rock and melting it. A fulgurite itself is agnostic as to the origin of the spark. In most cases, the spark is from lightning. However, enterprising individuals are able to make their own artificial fulgurites from sand purchased at the local hardware store, powerful electric equipment (not purchased from the local hardware store), and a willingness to shock themselves silly with high power electric equipment. Objects made by such a method are still fulgurites. There are subtle differences in shape, form, color, and composition that allow discernment from lightning fulgurites, but in general, it's hard to differentiate these objects from lightning-formed fulgurites. That said, given the relatively large overhead cost of the equipment needed to make fulgurites, the high price of power for making these objects, and the comparatively low value of fulgurites, "counterfeit" fulgurites aren't really going to be common on the market. I wouldn't really worry about those too much. Most sellers are honest enough to advertise them as artificial fulgurites, as they typically carry a premium over natural fulgurites due to the rather large expense for making them.

So what can help separate a fulgurite from a non-fulgurite? The first, and easiest way of separating them is by the tubular morphology. In fact, few people will claim non-fulgurites are fulgurites if they don't have the tubular shape, since this is such an obvious characteristic. Most fulgurites are indeed tubes (it should be noted here that some fulgurites aren't tubes, and instead are glassy melts formed on the surface of rocks or droplets, but both occurrences are relatively rare and generally unrecognized). Much more important for identifying a rock as a fulgurite is the presence of glass. Glass, if present, should be fairly obvious- it should be shiny and have a vitreous luster. Luster is the how light shines off of a surface, and "vitreous luster" means the light shines off of it as it would from glass. So it should look shiny like glass. Additionally, if the glass is all fractured, it should look like shards of glass with conchoidal fractures (like the edge of newly-broken glass).

A common non-fulgurite that is sold as a fulgurite is a filled burrow. A burrow, not burro, is a hole dug by a creature. Many creatures make burrows, from well-known ones such as prairie dogs, moles, and earthworms to less obvious ones such as owls and turtles. Many insects make burrows as well. If you throw marine life into the mix, then crabs, shrimp, some fish, many mollusks, and a plethora of worms also make burrows. Burrows are made in soil on land and in the sand in marine environments. In the case of marine organisms, the burrows stay put and don't collapse because the animals that make them do so by lining the hole with mucus. An unexpected effect of animals making burrows is that the burrows can actually be stronger than the surrounding soil and sediment when that soil and sediment turn into rock. As a result, a burrow may stick out of the rock as the surrounding rock weathers away. The result is a tube that is usually made out of sand, that kind of looks like a fulgurite.

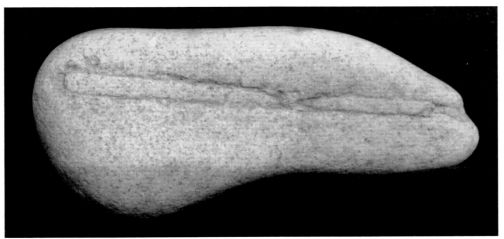

A fossilized worm tube, *skolithos*.[13] Width of the whole rock is about 4 inches (10 cm).

The key distinguishing feature here is then the presence of glass. If you find a tube on a beach, look for glass. If there's no glass, it's probably a fossilized burrow. While neat in their own right, they aren't fulgurites.

The second most common non-fulgurite is an iron oxide encrustation that formed around a tree root. Trees, like people, breathe oxygen from the air. At times, this oxygen will percolate down to the roots of a tree. When this happens, the oxygen can get released into the ground. If the ground is wet, and the conditions are right, then this oxygen can react with iron that is in the soil to produce a layer of rust that surrounds the root. When this root, or the entire tree dies, the rusty remnant remains as rock. When the wood rots out of this material, it can produce a tubular shape when excavated that is hollow. Such an iron oxide material can be quite reminiscent of a fulgurite.

---

[13] Skolithos are interesting fossils in and of themselves. The actual creature that made them is not known, but was likely worm-like. Skolithos are especially important fossils in the state of Virginia. They are diagnostic of the Chilhowee Group, the oldest Cambrian rocks found in Virginia (about 540 million years old). As a young child, I once found a skolithos fossil on a beach in Virginia, and brought it to my 1st grade teacher. She identified it as a petrified stick. It took fifteen years and a degree in geology to figure out what it really was.

Iron oxide root tube.

Again the discerning factor here between tree root rust tubes and fulgurites is the lack of glass. Tree root rust tubes do not have glass, whereas fulgurites do. Additionally, rust is not stable when it is struck by lightning. Instead, the rust bakes and fuses to form a gray mineral called hematite, or even to form a black mineral called magnetite. This comes as a natural consequence of the heat driving off excess water, and turning the material into rock.

There are other non-fulgurites out there, but these are the two the most commonly encountered. In general, the glass rule will work well in distinguishing a fulgurite from a non-fulgurite, with the exception of a few obscure varieties of fulgurite that happen to occur in calcium-rich sand or soil. Glass is still present in these fulgurites, but it's microscopic in this case.

Additionally, not all glass is from a fulgurite. Beach glass is most likely from old bottles, as fulgurites don't survive tumbling in the ocean too well. I've had one example of a giant non-fulgurite brought in for my inspection that was likely a manmade tube of glass. Where it came from and why it got there, I don't know, but it wasn't a fulgurite. The tube rule works well for differentiating fulgurites from glassy non-fulgurites.

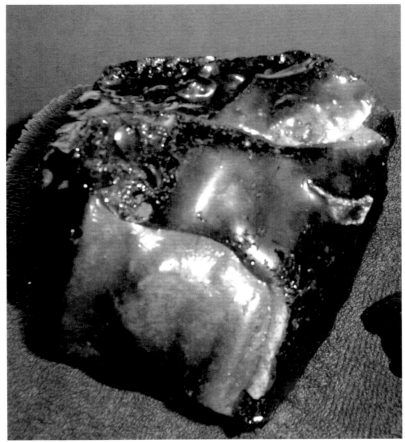
Giant, glassy non-fulgurite of unknown origin.

10. Fulgurites in Pop Culture

One of the reasons most people think the best places to find fulgurites is at the beach is because of the 2002 romantic comedy *Sweet Home Alabama*. In this movie, the male love interest makes fulgurites by driving metallic rods into the beach during storms and excavating the fulgurites. The fulgurites are intricate globs of pure, clear glass that loop in unusual patterns, and are sold at an art studio. In short, they look nothing at all like a fulgurite. However, the movie did very well at the box office, and many people believe the fulgurites portrayed there to be accurate representations of nature. This is also the source of the assumption that fulgurites are easy to find, and make, at the beach.

On the other end of the movie spectrum was *The Reaping*. In this movie, fulgurites are featured only at the beginning of the story, which involves a skeptic traveling to Louisiana to investigate paranormal phenomena. These fulgurites are used by the locals to make natural wind chimes. In *The Reaping*, instead of getting a message of "love the one you're with" as in *Sweet Home Alabama*, the message is actually "first-born devil worshippers are trying to bring on the Apocalypse". Fulgurite wind chimes could potentially work, though using sand fulgurites would most likely result in broken shards of glass after the first strong wind gust.

As far as I know, those are the only two movies that have ever featured fulgurites (as much as rocks ever get featured in film). Two other television shows also had fulgurites featured: *Supernatural* and *The X-Files*.

The older of the two, *The X-Files*, had an episode with a teenager who was able to control electricity. He uses his power to resuscitate truck drivers, change the channels of the TV without a remote, and fry some cows. The latter activity occurs when the teenager summons lightning down in a pasture, which strikes him and the nearby bovines. Intriguingly, a fulgurite is left over, with two sneaker footprints preserved within the glass. This episode of the X-Files actually had the most accurate fulgurite shown by Hollywood, pleasantly enough. It was gray, with green-black glass in the interior, matching a type 2 fulgurite quite well.[14]

The last major fulgurite feature episode is of *Supernatural*. In this episode, the two main characters, brothers Sam and Dean, use a fulgurite to bind Death, to kill an angel who has become a power-hungry god. I'm not sure I can follow that up with a joke, honestly. A fulgurite is phrased as an "Act of God Crystallized Forever", incorrectly using the word crystallized. Sam and Dean steal the fulgurite from a collector as they believe fulgurites to be quite rare. The prop itself is a lumpy gray tube, which might pass as a fulgurite if not looked at for too long. That said, in a following episode, they are told by other demon hunters that they can get fulgurites from new age crystal shops, suggesting the show's producers recognized, or were told that fulgurites really aren't that rare.[15]

In summary, the *X-Files* gets the highest props for having a good and fairly accurate fulgurite, whereas most other Hollywood manifestations of fulgurites are closer to "Hollywood magic" than to reality.

---

[14] The episode here is "D.P.O.", short for "Darin Peter Oswald", the boy with the lightning powers. It's the third episode of season 3. A young Jack Black is a co-star. https://en.wikipedia.org/wiki/D.P.O._(The_X-Files)

[15] This episode is "Meet the New Boss", referring to the replacing of God with one of his angels. It is the first episode of season 7. You can see more here:
http://www.supernaturalwiki.com/index.php?title=7.01_Meet_the_New_Boss

11. Fulgurites: the Science

The science of fulgurites was pretty thin after the 19th century. In large part this was because their coolest property had been figured out. Lightning melts soil, end of discussion. That alone is pretty neat, and, to most geologists, there wasn't a lot else that needed study. In the recent body of scientific work, there were a few major works highlighting some of the weirder features of fulgurites, including Eric Essene's study of the Winans lake fulgurite, and Terry Daly and Peter Buseck's work as part of the study of the formation of fullerenes in fulgurites. Scientific papers on fulgurites might pop up once every few years, a relatively slow rate until the 21st century. In part this was due to the advancement of key geological and chemical tools that made the analysis of these rocks easier and faster. Prior to that, finding interesting things took more effort and time.

In the early 2000s, Dr. Martin Uman, at the University of Florida in Gainesville, FL, led with colleagues a series of experiments to understand some of the fundamental features of lightning. In these experiments, the research team launched model rockets at summer thunderstorms. These model rockets were trailed by a copper wire, which provided the conductor necessary for lightning to flow. From these experiments, the researchers hoped to get key data on lightning so that they could understand some of the physical features of lightning and how it forms. It turned out the copper wire may have not been necessary and instead the exhaust from the model rocket may have provided enough ionization to trigger the lightning.[16]

One of the side projects of this work was some investigation of fulgurite formation. Fulgurites were formed using these rockets by tying the copper line to a batch of sand. The lightning, triggered by the model rocket, would then penetrate the sand and form a fulgurite. Using this method, Dr. Uman with collaborator Kevin Jones and his daughter formed fulgurites in several materials that would normally never be struck by lightning, including copper oxide, nickel oxide, and manganese oxide. These would never be struck by lightning principally because they're never found in abundant enough a concentration to make a rock that could be struck. However, this work revealed several insights into fulgurite formation, including mineral changes. These mineral changes were loss of oxygen and formation of metal, and paralleled work done by Essene and Fisher back in the 1980s. This work suggested that fulgurites could form some metals through the heating process, though they could not make metallic iron or silicon this way.

Another major project, coming to fruition in the mid-2000s, was a study of climate history preserved within fulgurites. Led by Rafael Navarro-Gonzalez from the Universidad Nacional Autónoma de México, the researchers analyzed a fulgurite from the Sahara, specifically focusing on trapped bubbles frozen within the glass of the fulgurite. When the trapped gases within these bubbles were analyzed, the researchers discovered that the environment where this fulgurite was formed in was wetter and had more plants than present. When the fulgurite were dated using geologic techniques (more on that later), the implication became clearer: the Sahara desert wasn't nearly as desert-y as it is now, and the climate within the region had only changed to dry and hot in the last 15,000 years. While that had been known at least in part before, it was

---

[16] You can see the results of the experiments at the team's website, as well as good pictures of what goes on there: http://www.lightning.ece.ufl.edu/

the tool used to answer this question that made the research novel. Fulgurites finally had found a use in geology besides just being weird rocks.[17]

My own story starts around here. I have been active in fulgurite research for about ten years. At present, I have written or been an author on a plurality of scientific publications on fulgurites in recent history. This is not to say a lot- I have six articles on these objects, with the next most prolific authors having two. So, it is not a highly active field of research, when other fields of study can get thousands of papers in a single year. It is also only a side project of the main research focus of my own work, but because fulgurites have been so poorly studied, any work done on these objects can advance our knowledge of these unique rocks.

My own research began when I was a close-to-graduating grad student at the University of Arizona. At Arizona, I worked with planetary scientists who were quite interested in impact craters. Notably professor Jay Melosh, who is viewed by many as the world's expert in impact cratering physics (recognized, for instance, by the US National Academy of Science), was working with other scientists to see if there was a connection between lightning and impact crater glasses. My own research centered on the chemistry of the element phosphorus when it's in space and on the earth. I talked frequently with Jay Melosh's grad student, Abby Sheffer, on her work on fulgurites. Abby researched the chemistry of iron in fulgurites. Her work, which was eventually published as her dissertation, concerned how iron changed its redox state when heated by lightning.[18]

A brief chemistry lesson/reminder. Chemistry is often concerned with the *oxidation state* of materials. The oxidation state is a count of the number of electrons around an element compared to the number of electrons that should be around the element. For instance, the element oxygen has a total of eight electrons in the atomic form. This is equal to its atomic number, which is also eight. The air you breathe has oxygen with nominally eight electrons around it. Because the difference between the expected number and the actual number is zero, the oxidation state of oxygen in air is also zero. Normally, oxygen is quite reactive, and will readily react with other elements to steal their electrons. For instance, in water (H_2O), the oxygen hogs both of the electrons that are normally around the two hydrogen atoms, and hence it has a total of ten electrons, as opposed to the eight it normally. Since there's an excess of two electrons, and since electrons have negative charge, the oxidation state is said to be minus two (-2).[19] Each of the hydrogen atoms in water is presumed to have an oxidation state of plus one (+1), to balance the charge of the oxygen.

Abby's research investigated the difference in oxidation state of iron in fulgurites. Normally, iron has an oxidation state of +3. This is the iron that is familiar to us as rust. Iron, when it's in contact with air, loses its electrons readily to oxygen, in a process called *oxidation*. This is the eventual fate of iron metal sitting on the earth's surface. In contrast, when Abby investigated iron oxidation state of fulgurites, she found this iron was reduced in oxidation state, turning to +2, and at times, even iron metal (nominally a zero oxidation state). It had gained electrons, a process called *reduction*.

---

[17] The original paper is here: http://geology.geoscienceworld.org/content/35/2/171, and you can find a summary of the article here: http://www.sciencemag.org/news/2007/02/spark-sand

[18] Abby's dissertation is here, in all of its 46 MB glory: http://arizona.openrepository.com/arizona/handle/10150/194729

[19] Benjamin Franklin is to blame for the confusion of having too many electrons equating to being negative. In his early experiments, he saw that one material transferred charge to the other when static electricity had built up, and by unfortunate accident, termed the one with more electrons "negative".

In many ways, this research paralleled the findings of Essene and Fisher. They, too, had found iron metal in fulgurites. The difference was that Abby's work looked at several different fulgurites, from a variety of locations. It was apparent that this reduction was broader than the smelter-style reduction shown in the Winans Lake fulgurite, and occurred in more target materials that sandy soil with tree roots in it. Abby's study was one of the first to investigate a suite of fulgurites, and had provided the first glimpse of the changes to chemistry that result when lightning strikes rocks, sand and soil.

When I graduated with my Ph.D., I was prompted by Abby's work and by a modest bump in salary (the ramen became gourmet ramen), I bought a set of fulgurites, primarily from eBay, and began to investigate the geochemical changes that occur when lightning strikes soil. I was particularly interested in the chemistry of phosphorus. However, the set of fulgurites I purchased varied significantly in form and shape, implying that how each fulgurite forms is dependent on the material it forms in. Additionally, I handed off a few of the samples of fulgurites I procured to some colleagues and they (a few years later) came back with a lot of unusual and potentially interesting geochemistry reported. This research started at the University of Arizona, but continued when I move to the University of South Florida to take up a job as an assistant professor in the geology department. Following up with this work, I sampled several fulgurites from Florida, and came to realize they revealed something intriguing about the fundamental nature of lightning.

The next few sections of this book will discuss the scientific impact of fulgurites. In general, I will try to avoid using too much jargon, and will also avoid any more formulae and equations.

12. What types of fulgurites are there?

Any suitably large and varied collection of fulgurites will readily reveal that fulgurites come in a few distinctive shapes and sizes. They also come in a variety of different colors. The reasons for these differences are two-fold. The first is target material differences. Sand is physically different from rock, which is different from clayey soil. They differ in terms of their composition, the size of individual particles and grains, the strength of the material, the density of the material, and the amount of water that may be present within the spaces between grains. They may also differ in terms of the amount of vegetation present, which may change how lightning travels through the target material, and whether or not the carbon from the tree reacts with minerals on ignition.

The second difference is in the properties of the lightning strike that formed the fulgurite. Lightning, as we will show, varies significantly between individual strikes. Some strikes will kill you outright, others will merely maim you. Lightning varies in terms of the energy of the lightning strike, the timing of the lightning strike, the number of strokes, and even the electric charge that is transferred. Several of these properties may vary individually by factors of 100. That is, the largest value divided by the smallest value of an individual property (for instance, energy), may be as great or greater than 100. For energy, we have actually found the difference to be a factor of almost 300.

As a result, there is necessarily a significant variation between fulgurites. Sometimes low-energy lightning strikes sand, sometimes high-energy lightning strikes the top of a mountain. The result is that fulgurites, even when formed in the same target material, can vary significantly in size, shape, distribution, and even color.

Along with compositional differences, the differences between lightning strikes are the major source of variation between fulgurites. The results of these differences have been known for some time. James Richardson, talking to mineral enthusiasts at a local Philadelphia mineral club in the 1890s, described fulgurites to his club in a lecture that was preserved in the meeting annals. He argued at that time that there were two types of fulgurites: tubes and encrustations or globs.[20]

This division works well enough even today. There are two types of lightning-formed glass: those that provide a history of the direction that lightning traveled, and those that don't. The tubular fulgurites are the former, and other varieties are the latter. It is many times easier to identify the tubular variety, especially since non-tubular fulgurites overlap with other varieties of glass, with respect to their shape.

The two forms are often related: glassy globular fulgurites are often directly associated with tubular fulgurites. Additionally, in the few places where encrustation-style fulgurites are found, tubular fulgurites will usually also penetrate the encrusted rock. The best identification of a fulgurite is, again, the presence of glass within or on the rock.

In some ways, this simple division between tube and non-tube fulgurites persists even today. Two articles appearing in the rock enthusiast magazine *Rocks & Minerals* describe a novel variety of fulgurite, termed "Exogenic fulgurite". The first, by J.W. Mohling, describes some strange glassy splash-form fulgurites coming from Elko, Nevada. Seven years later an article written by Michael Walter describes the occurrence of a new group of exogenic fulgurites from Oswego, New York. These fulgurites, which are droplet, or splash-form in shape, were

---

[20] You can read the original discussion in "The Mineral Collector", volume 4, from 1898. A few other discussions also pop up in this journal, also by Richardson.

formed as an accompaniment of tubular fulgurites. The term exogenic means the fulgurites were formed (or found, rather) outside of the main tube of the fulgurite. They form as droplets of molten glass sputter out of a fulgurite tube after lightning strikes.[21]

An exogenic fulgurite from Tucson, AZ. The main droplet width is about 1.5 cm.

Detailed study of fulgurites has revealed that variations between fulgurites are best attributed to differences in composition than to differences in morphology or shape. In other words, the material that makes up a fulgurite is a lot more defining of the fulgurite than the shape of the rock. For one, encrustation-style and globular fulgurites are never found where fulgurites formed in sand are found. Secondly, fulgurites formed in sand are very thin whereas fulgurites formed in clay or rock or non-sand soil tended to be much thicker. Third, a new variety of fulgurite was found to occur in carbonate-rich soil, and these fulgurites tended to be much thicker still, with little to no void whatsoever.

Thus, fulgurites could be grouped quite well into four separate groups or types. We proposed these four types were sand fulgurites, soil/clay fulgurites, calcic-soil fulgurites, and

---

[21] For more details, check out two articles in the rock collecting journal, "Rocks and Minerals". J.W. Morling (2004) Exogenic fulgurites from Elko County, Nevada: a new class of fulgurite associated with large soil-gravel fulgurite tubes. Rocks & Minerals, 79, issue 5. And M. Walter (2011) An exogenic fulugirte occurrence in Oswego, Oswego County, New York. Rocks & Minerals, 86, issue 3.

rock fulgurites. This division works primarily for natural fulgurites, or in other words, those fulgurites formed by lightning in naturally-occurring rocks and soils. Artificial or anthropogenic fulgurites introduce much more variation in the type of material that gets struck (for instance, a sidewalk, or a metal pole stuck into the ground), and may also vary with the duration of energy that is deposited into the ground (a downed power line may discharge for hours into soil).

Fulgurites appeared to group nicely into several distinct classes without too much hassle. The Type I fulgurites were assigned to sand fulgurites. These fulgurites are characterized by very thin glass walls, typically about 1 millimeter thick, and large void spaces with respect to the total size of the fulgurite. These are the fulgurites most people tend to think of when they think of fulgurites. They are formed in sand, often on dunes nearby beaches (but not necessarily on beaches). These objects have a low density much akin to that of the sand they were formed in. They do have some variety in color, and are typically gray or brown (especially in iron-stained soil), and a few come in as white as pure quartz sand.

Type I fulgurite from Polk County, Florida. Void width is 1 cm.

The Type II fulgurites are, in my opinion the most interesting group of fulgurites. These fulgurites formed in soil. They were originally clay, or fine sand that was not just quartz, and often have small pebbles embedded on their exterior. They usually have a gray exterior, characteristic of the frying or fusing of minerals on their outer surface. This gray crust turns into a distinctive glass as you go inward into the fulgurite, and usually hits the central void space thereafter. These fulgurites can be thick, up to a few inches in diameter, although typically they'll be about an inch (or between 2-3 cm). They are relatively rarer than the type I fulgurites, since the Type I fulgurites are easier to find as they stick out of loose sand. Also, since they are made of a glass that is less stable than quartz glass (made of SiO_2), they don't last as long as Type I fulgurites.

Type II fulgurites can have many different colors, sizes, and shapes. Type II fulgurites often branch and will have small vents that feed off of a larger, central vent. The colors of the glass of these fulgurites range from brown-black to green to an occasional red or even a deep sky blue.

Type II fulgurite from Western New York. Width is 10 cm.

     The Type III fulgurites are currently the least well-studied. We first reported these fulgurites in a paper in 2009. Tubes of tan-brown colored material were found in the American Southwest. These were composed primarily of calcite and quartz sand. When we first bought them, we did so principally because they were advertised on eBay as being fulgurites, and, not knowing much about fulgurites at the time, we were taking a gamble. It turns out that they were indeed fulgurites, but unlike the others which are obviously fulgurites based on the presence of glass, these were composed mostly of a calcium carbonate mineral known as calcite. Globally, calcite is an important mineral as it's the major component of the rock limestone, which is very common across large swaths of the continents. Here, too, calcite played a role, as the soil in the American southwest is composed often enough of something called desert hardpan or caliche. This material, which makes soil hard as rock, is common to desert environments. In the summer in places like Arizona, monsoon thunderstorms drop half of the year's rain in less than two months, and these events are well-known for frequent and dangerous lightning. This lightning, like lightning everywhere else, strikes the caliche, frying it into a tube shape. These fulgurites tend to be beige in color, have little no central void, or a void that has been filled in with mineral grains in the years that followed. The glass tends to be adjacent to the fulgurite center, and is not necessarily obvious.

Type III fulgurite from Sweetwater, Wyoming. Width is 1.5 cm.

The final group of fulgurites is one that has been studied fairly well. These fulgurites occur in or on rock, and are called Type IV. They include the "encrustation" variety of fulgurite, often forming a black glass on top of the surface of a rock. It sometimes looks like lichen, and may not be too obvious when encountered. They typically occur on hill or mountain tops, on exposed rock faces. In general, the encrustation variety forms a black glass glaze on the rock surface, and the tube variety forms small tubes that penetrate the rock. The tubular variety tends to be fairly narrow, as rock is more solid than clay or sand, and the fulgurite is denser when it occurs in rock.

Type IV fulgurite glass from Kaleetan Mountain, WA, USA. White arrows note some of the fulgurite glass, which is black and red/brown.

By and large, the rock/sand fulgurite dichotomy is the most common division seen when talking about fulgurites. I do note, though, that Wikipedia has divided fulgurites into the four type scheme above, following my own research. I didn't even edit the article, so this change looks to have been a happy consequence of being accepted in the scientific literature. I tell my classes that "My research seeks to understand the nature of lightning and has been published in journals such as Nature Geoscience and Scientific Reports," speaking through their yawns and indifference. Then I mention that the work has even been picked up by Wikipedia, and I have their attention yet again![22]

---

[22] https://en.wikipedia.org/wiki/Fulgurite, see the classification section.

13. Weird Fulgurite Minerals and the Rule of Threes

Fulgurites have experienced a small-scale, but distinguishable, scientific renaissance. In the decades preceding the 2000s, a scientific paper on fulgurites would pop up maybe once or twice a decade. They were not studied often, and when they were, it was often because of boredom on the part of the scientist, or due to bright-eyed and bushy-tailed young scientists looking to check out some new rocks.

The turning point in the literature on fulgurites occurred with the publishing of Essene and Fisher's study of the Winans Lake fulgurite. This was the first paper that showed, "Hey look, these fulgurites can be kind of weird… maybe we should study them some more!". Although it took a few decades to make more headwind on the subject, this discovery has helped fuel modern fulgurite research, specifically on the formation of weird minerals within the masses of these objects.

The most unusual of these minerals contained elemental silicon. Now silicon may be familiar to you as the stuff in those packets found in beef jerky that you really, really shouldn't eat, or in the implants placed by cosmetic surgeons, but in both cases that's actually silicone, with an extra "e". Silicon is the element. It is not a particularly rare element, as these things go. In fact, on the surface of the earth, it is the second most common element, right after oxygen. This is because most of the rocks and minerals on the surface of the earth are made of silica, either as a whole or as a portion, which is one silicon atom bound proportionately to two oxygen atoms. So finding silicon in and of itself may not seem to be too interesting.

On the earth, nearly all silicon is bound to oxygen. Finding silicon in a form not bound to oxygen is quite uncommon. Most of the times silicon has been found *in silico solo* have actually been in rocks from outer space. A discovery of elemental silicon in a fulgurite places these rocks in the same cadre as meteorites. And with silicon's co-occurrence with metallic iron (which is weird again because it's not stable on the surface of the earth, and is typically found only in meteorites), that's two distinct marks for fulgurites being strange and worth taking a second look at.

Reminder

- A mineral is a **naturally occurring, ordered solid**, generally consisting of a **defined composition** and generally **inorganic**
  - So no man-made materials here
  - It can't be random in structure- no coal
  - No ultracooled liquids allowed
  - **Must not vary much between different occurrences**
  - **Should not be carbon-based (though exceptions exist to this rule)- so sugar isn't a mineral**

Geologists have a weird little secret: most rocks are actually kind of boring. Once a rock is characterized, it's usually easy to place the rock in a set of categories that tell us most of the geological information we care to know. There may be a few tidbits of coolness to every rock, but usually the general information is easily understood. As a result, when things are weird, geologists tend to take notice. If a rock that is formed by lightning has the same minerals as those that fall from space, then that's worth noticing.

From where then did these unusual minerals come? There's two options: first, they were present in the original target, or second, they were formed by lightning. The first is fairly readily eliminated. If we look at any given sample of soil, there will not be any elemental silicon (99.99% guaranteed), outside of a few spots where it might have been dumped in as industrial waste. Winans Lake was not such a place. These minerals are also not stable over the long-term, rusting and oxidizing to produce the minerals that are more common on the earth's surface.

We've talked a fair bit about the minerals that were found in the Winans Lake fulgurite. Why then did these minerals catch the attention of Eric Essene and Daniel Fisher so much? A reminder on the definition of a mineral. A mineral is a solid substance, with a specific chemical composition or range of compositions, and a specific chemical structure, which occurs naturally on the surface of the earth or of other planetary bodies. Let's break that down a bit. First, a mineral must be a solid. Liquids and gases get excluded there. Technically, that may also exclude some glasses, since glasses are super-cooled, slow-flowing liquids. Next, they have a specific chemical composition. This allows us to differentiate between minerals that are similar in structure but differ in what they're made of. The corollary of this concerns structure. A mineral has a specific structure, and even if it has the same composition, it may be a different mineral. The best case is raw carbon. This can be either graphite or, more pleasingly, diamond. Finally, it needs to occur naturally on the surface of the earth or other planetary body (including an asteroid). A chemical that comes out of a chemistry lab is unnatural, and is probably not a mineral in that case. A mineral is usually inorganic, but is not required to be so. There are some strange minerals that are in fact at least partially organic, such as the mineral whewellite, a calcium oxalate salt. Oxalate is a metabolic by-product of several reactions that take place in living organisms, the oxalate molecule also contains carbon.

The Winans Lake fulgurite found a close companion in the York County Fulgurite, from Pennsylvania, USA. This fulgurite, which we reported on in a few of our fulgurite papers, had expelled from its tube a droplet fulgurite (exogenic fulgurite) that, when cut open, had a large metal grain inside of it. This metal grain had minerals quite similar to the Winans Lake fulgurite. Although I say "large", the word "large" is in the micro-sense. They were about a millimeter wide, a bit larger than the period at the end of this sentence. Nonetheless, with modern microscope techniques, we could see plenty of features in these drops.

The minerals found in the Winans Lake fulgurite were composed of mixtures of roughly four elements: iron and titanium, and silicon and phosphorus. The York Pennsylvania fulgurite added two more elements to these four: nickel and sulfur. We can divide these elements into metals and nonmetals. The metal elements—iron, nickel, and titanium—were an important half of the mineralogy of fulgurite metals. The other three elements—silicon, phosphorus, and sulfur—non-metals all, formed the counterbalance to these metals.

These metals are not a common occurrence in fulgurites. There has only been a handful of minerals like these reported in fulgurites. Abby Sheffer lists one from the Cacopon River, West Virginia that contains minerals similar to those of Winans Lake, and there's been a report of similar minerals in a fulgurite from Hidalgo, Mexico. Metals were found in only one of Abby's fulgurites (about 10 total), and we found metal in only one of nine of our samples. The conditions necessary to form these minerals appears to be rather rare, even for fulgurites, and maybe 10% of fulgurites will have these minerals on a large scale.

We call these minerals iron silicides, though some may have no iron and others no silicon. However, most of these minerals were very specific combinations of metals to non-metals. For instance, a mineral with the formula FeSi was found in all three fulgurites from Winans Lake, West Virginia, and York Pennsylvania.[23] So each set of fulgurites shared this in common. However, the Winans Lake and the West Virginia fulgurites differed from the York Pennsylvania fulgurite in that the other minerals were non-metal rich. One mineral, with a formula Fe_3Si_7, a ratio of 3:7 iron to silicon, was found in the Winans Lake fulgurite, and a $FeSi_2$ mineral (1:2 iron to silicon) was found in the West Virginia fulgurite. The York Pennsylvania fulgurite in contrast had a suite of minerals that were metal-rich. We noted about six different variations (see if you can spot the theme): $FeSi$, Fe_5Si_3, Fe_2Si, Fe_7Si_3, Fe_8Si_3, and Fe_3Si. If the pattern is not apparent, try multiplying some of those by three to get three silicon. These then become Fe_3Si_3, Fe_5Si_3, Fe_6Si_3, Fe_7Si_3, Fe_8Si_3, and Fe_9Si_3. With the exception of the missing Fe_4Si_3, we found distinct ratios between iron and silicon. The Winans Lake fulgurite and West Virginia fulgurite likely also have the same feature, but in reverse.

---

[23] Our paper on this fulgurite, and several others, can be found here:
https://link.springer.com/article/10.1007/s00410-012-0753-5

In general, titanium can substitute in for iron in these minerals, and phosphorus substitutes for silicon. Although rarer, nickel and manganese can sub for iron as well, and sulfur and potentially aluminum can mix with silicon. However, the rule of three holds well for these unusual minerals, when you sum up the metals and the nonmetals.

14. The special case of phosphorus

The element phosphorus received some mention in the above part on weird minerals in fulgurites. This is because in large part my work on fulgurites got its start with looking at phosphorus chemistry.

As a mineral constituent phosphorus is one of the more boring elements in geology. Like the element silicon and silicate minerals, phosphorus is generally recognized as only one form: as phosphate minerals. My work has focused on the development of life on the early earth, and phosphorus is a big part of that. Phosphorus is an important element in biology as it forms the backbone of DNA, and is the main active part of the energy transfer molecule ATP. Part of my research concerns the spontaneous formation of phosphate-biomolecules, and an operating hypothesis within this field is that meteorite minerals, including a mineral known as schreibersite, may have provided the phosphorus for the origin of these molecules.[24] This mineral, which has a formula of Fe_3P with nickel sometimes substituting for iron, may ring a bell with the "rule of three" above. Indeed, the mineral schreibersite is one that is found in fulgurites and sparingly few other places on the surface of the earth.

Fulgurites have schreibersite for the same reason they have Fe_3Si: extreme reduction, probably promoted by the burning of organic matter. That we found schreibersite within the York County Fulgurite, and that it was also found in the Winans Lake fulgurite shows that the extreme reduction process that occurs in fulgurites is one of the few ways on the earth's surface that this mineral can get made. However, in our analysis of several different fulgurites, we found only the one from York County happened to have schreibersite.

That said, there were some other unusual results that came out of this study. We analyzed several Type III fulgurites and discovered that these two had phosphorus that was not in the form of phosphate (P with an oxidation state of 5+), but in a form not typically found on the surface of the earth, phosphite (P with an oxidation state of 3+). That change in vowel may not seem too significant, but has in fact been an important discovery that has some far-reaching implications in biology, sustainability, and the future of humanity. While the fulgurites weren't key to that, they do hold their place in the discovery of the first form on not-phosphate and not-schreibersite phosphorus.[25]

Our goal initially with this discovery was to determine where the phosphite was within the fulgurites. We identified the source within the York fulgurite to be the schreibersite, but the source within the Type III fulgurites was less obvious. There were no metals whatsoever within these fulgurites. From where then did this phosphite come?

Kristin Block, a grad student working with me at the time, came in with excitement after spending the day in the cold basement housing an electron microprobe in our home building at the University of Arizona. She had found the probable reduced P carrier in a type III fulgurite from Yuma County, Arizona. It was a strange mineral grain, with a melted, mottled pattern, consisting of half calcite, and the other half a calcium-phosphorus phase. Unfortunately, the grain was too small and warped to get an accurate measurement on its composition. However,

---

[24] You can find the paper presenting this hypothesis here:
http://online.liebertpub.com/doi/abs/10.1089/ast.2005.5.515 Also, some summaries can be found here:
http://www.astrobio.net/meteoritescomets-and-asteroids/meteorite-phosphorus-aided-life-on-early-earth/

[25] This discovery was presented here: http://www.nature.com/ngeo/journal/v2/n8/full/ngeo580.html

with the other data from the York County, Pennsylvania fulgurite, which was big enough to get an accurate measurement on, we had enough data to publish a paper.

So what does this mean for the element phosphorus? It turns out that microbes, of many varieties, are capable of using phosphite as a nutrient, instead of phosphate. Most organisms, including you and me, must use phosphate, but microbes tend to be much more versatile, and hence some species can use phosphite. Prior to this work, there were no known natural sources of phosphite. People have been making phosphite since the middle ages, but the microbes that are able to use phosphite seem to be much more ancient than that.

Sources of this phosphite may have been meteorites, though there's probably not enough from those to really promote microbial change. Lightning-produced phosphite adds another source of this material. How much phosphite might lightning actually provide?

Fortunately, a simple set of multiplications of known or somewhat known values (called parameters) can provide an answer. We can use an equation:

$$P = f_{lightning} \times CG_{fulgurite} \times P_{soil} \times \%_{reduced} \times M_F$$

Each of these values is multiplied together to give us an estimate of the amount of phosphite produced each year, or "P" above. The parameter $f_{lightning}$, is the frequency of lightning across the surface of the earth. We know this value fairly well now from satellite observations. It's about 45 flashes every second (over a billion flashes per year). That may seem to be a lot, but the earth is pretty big, and this translates to only about a few thunderstorms active anywhere on the surface of the earth. The next value is $CG_{fulgurite}$, the fraction of this lightning that is cloud-to-ground and forms a fulgurite. The first part of this value, the fraction of lightning that is cloud-to-ground, is known to be about 1 in 4 (as we mentioned in chapter 7). The second part is less clear, but is probably not far from 100%. If we multiply ¼ by 90%, we get about 0.23.

The next values deal with the fraction of phosphorus that is changed by lightning when a fulgurite is formed. P_{soil} is the amount of phosphorus that is in soil, or about 750 parts per million on average. A part per million is similar to a part per hundred, which is better known as a percent. Just as 15% is equivalent to 0.15 (15 divided by 100), a value of 750 parts per million translates to 0.00075 (750 divided by one million). The following two parameters reflect the amount of phosphorus that gets reduced by lightning on average ($\%_{reduced}$), a value that is hard to estimate accurately given how few measurements have been made of it. A value of 20% is likely, or 0.2. The final value is the average mass of a fulgurite, M_F. We don't know this value too well as a typical mass of fulgurite is between a gram and a kilogram, but if we take a typical value between these of 30 g (about an ounce), we can solve for the amount of phosphorus reduced by lightning each year.

Multiply these values together, and we arrive at about 1000 kg, or a ton, each year. That's not really a lot when you consider the entire size of the earth. Still, it is a consistent source of phosphite that may provide a low, background concentration of an important element.

15. Reduction in Fulgurites

These unusual phosphorus, silicon, and iron minerals had to come, newly formed, from the lightning strike. We've mentioned that the likely reason these minerals were present in the fulgurites was carbon from plant matter vaporizing and pulling oxygen atoms off of mineral grains when heated by the lightning strike. Carbon reacts with iron oxides to pull off oxygen from iron to make iron metal, or with phosphates to make phosphorus, and silicates to make silicon, again with a parallel to a smelter changing ores to metals.

This does not address all possible routes to forming reduced material in lightning. Abby Sheffer's thesis work proposed that there was a general reduction of iron by lightning, in other words, transformation of iron 3+ (termed ferric iron) to iron 2+ (termed ferrous iron). This change is a lot smaller than the change from iron 3+ to iron zero, and it's quite likely there are some other routes to doing this as opposed to necessarily burning plant matter or tree roots to promote the reduction reaction.[26]

One possible route to transforming elements to a lower oxidation state is by reaction with atmospheric nitrogen. All lightning occurs at the surface of the earth, and the atmosphere of the earth at its surface is mostly nitrogen, N_2, making up 78% of the atmosphere. Oxygen, or O_2, comes in second at 20%, followed by argon at 1% and all the other gases, including CO_2 making up the last 1%. Depending on the environment, water as H_2O may also be present at a few percent, especially in high humidity climes.

Nitrogen, if it's within rock, may potentially react with oxygen in minerals to form nitrogen oxides and strip away O from minerals. Indeed, the reaction of nitrogen with oxygen is recognized as one of the few ways of actually turning atmospheric nitrogen into fertilizer, more specifically nitrate, which is a type of nitrogen oxide found in fertilizer. As such, lightning is important in "fixing" nitrogen so that life can use it. However, for this process to be useful in mineralogy, a lot of nitrogen would be required. Additionally, there's the issue that the reduction Abby saw in her fulgurites included several fulgurites that weren't very porous at all, meaning that there wasn't much air present nearby for this reaction to play any major role.

Likely a better route was reported by Jorg Schaller and colleagues in 2013.[27] This route is termed "Galvanic" reduction, named after Luigi Galvani, an Italian who helped develop the science of electricity and batteries. Galvani demonstrated, in typical 18th century scientist fashion, that electricity from a galvanic cell caused frog legs to spontaneously move. Fortunately, frogs and fulgurites are otherwise unrelated, and the case of galvanic reduction is the fact that electrons are transported by lightning into ground, reducing some materials as a result. This produces a net lower redox state for elements affected by lightning, as Schaller and colleagues showed through experiment.

---

[26] Reduction and oxidation are two processes that change the oxidation state of an element. They can be quite confusing. However, the origin of their names provides some background on why they are termed what they are. Centuries ago, chemists observed that when an element, such as a metal, was exposed to air, it would gain mass as oxygen reacted with it, *oxidizing* it. When the chemists altered this oxidized metal to remove the oxygen, the material was reduced in mass, experiencing *reduction*. Hence the origin of these names. These processes are combined to describe reactions in terms of reduction-oxidation, or redox.

[27] The paper is freely available here: https://www.nature.com/articles/srep03122

In case you were curious where the inspiration for *Frankenstein* came from…

While intriguing, the conditions explored by Schaller and colleagues were not accompanied by high-temperature changes that are associated with lightning. It is likely that these changes may be more important for the redox changes actually found within fulgurites.

When minerals are heated they can break down to form new minerals, a process we'll discuss in more detail later on. This is especially true for the iron oxide minerals. The iron oxide minerals include the mineral hematite (Fe_2O_3), and magnetite (Fe_3O_4), and a mineral not very familiar to most people, called wustite (FeO). When hematite is heated, it converts into magnetite whenever the amount of oxygen in the environment is low. It does so by releasing a small amount of O_2 gas. As a chemical reaction this would be represented as:

$3Fe_2O_3 = 2Fe_3O_4 + ½ O_2(gas)$

In my class, I have students calculate the oxidation state of iron in these two minerals. In hematite, the oxidation state is +3. In magnetite, the oxidation state confuses them, as it appears to be +8/3, which is slightly less than +3 (+9/3). In reality the oxidation state of the iron atoms is +3 for two of the iron atoms, and +2 for the other one. Heating thus causes a reduction of iron oxidation state. This is well-known in mineral chemistry, and can be represented by the graph below, called a mineral redox buffer diagram.

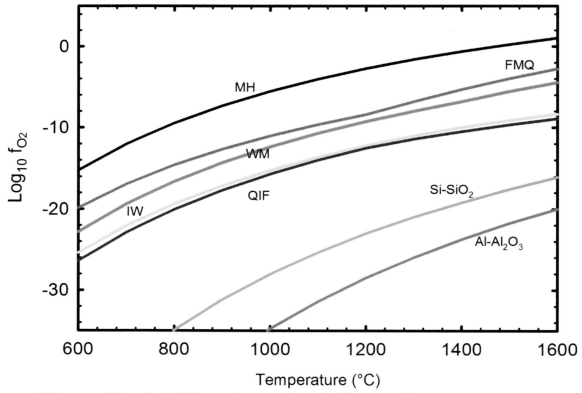

Ahh, one of the first full on sciencey charts! The x-axis is temperature, the y-axis is something called oxygen fugacity (roughly equal to pressure) of the gas O$_2$, on a logarithmic scale. The lines represent when certain minerals will dominate. Above the line, the more oxidizing mineral dominates, such as hematite in magnetite-hematite (MH), or wustite in wustite-magnetite (WM). Other abbreviations are for fayalite/magnetite-quartz (FMQ), iron-wustite (IW), quartz-iron/fayalite (QIF), silicon to silica, and aluminum to alumina. These are arranged from more oxidizing to more reducing.

This diagram, which shows the stability of minerals with respect to temperature in Celsius, shows that the more oxidizing minerals are less stable at higher temperature. They release oxygen under increasingly higher oxygen fugacities, or a measurement of how much O$_2$ gas is around. Oxygen fugacity is typically given in a logarithmic scale, meaning a value of -5 means there is 10^{-5} atmospheres (0.00001) of O$_2$, whereas a value of -20 means there is 10-20 atmospheres of O$_2$ (0.00000000000000000001). Big changes occur under relatively small changes to temperature.

Although reduction by heating could explain well the oxidation state change of iron when it was heated, and also explained Uman and Jones's discovery of reduction of copper and nickel oxides by lightning, it doesn't do as good a job of explaining the formation of elemental silicon or iron metal, or the reduction of phosphate. These latter cases require very low O_2 environmental conditions, unlikely to be present almost anywhere on the surface of the earth, and hence the presence of these minerals is still an open question in science. The best explanation is probably reduction by burning of carbon, but not all fulgurites appear to be associated with carbon sources.

16. Using Fulgurites to Calculate Lightning Energy

The most famous use of lightning in modern pop culture happened when lightning powered an ill-designed car from the early 1980s across the space-time continuum. In *Back to the Future*, since there's no place to acquire plutonium in the 1950s, Marty and Doc must rely instead on lightning striking a clock tower to power the Delorean. The precise power desired was 1.21 gigawatts, pronounced incorrectly as jigga-watts by Doc. Although the mispronunciation is just part of Hollywood, is the number anywhere near the actual ballpark for lightning? Just how powerful is lightning?

Fulgurites provide a means of measuring lightning power. But first, a quick discussion about what power actually means. Power is energy per time. Thus power is much like speed or velocity, which are distances per time. We've already talked a bit about energy. Energy is a way of measuring how much effort it would take to stop a moving object, how much heat an object can transfer to its surroundings, and the ability of an object to move up a hill. The common units of energy are calories, Calories, joules (and kilojoules and megajoules), ergs, kilowatt-hours, British thermal units or BTUs, and foot-pounds. You probably don't use too many of those in your day-to-day activities, and this is one of those cases that the Imperial unit (foot-pound) is actually used less than the SI unit (Joule), unless you happen to work in engines or firearms.

Power has comparatively fewer units, and the SI unit has really worked its way into American households. The two main units of power are the watt, which is on all sorts of appliances, and the horsepower, which is used primarily to describe motors in the US. Your light bulbs come in watts (W). A 60W light bulb is decently bright, and fills up a good corner of a room, whereas a 25W light bulb doesn't. A 100W light bulb probably hurts your eyes if you're close and stare at it. Your microwave is probably about 1000W, and you shouldn't stare at those. Your heaters have units of watts as well, often more than your microwave.

The watt may be somewhat familiar. Things that you can leave on have low wattage (such as a light or an alarm clock), whereas things that have high wattage you really shouldn't leave on (such as a convection heater) unless you're ready to pay for it. We can also measure engines in terms of wattage. A horsepower is about 750W. Thus a sports car, which may have a 600 horsepower engine, has a power of about 450,000 watts, or 450 kW.

How then does this relate to lightning and fulgurites? Fulgurites are "fossilized" lightning, in that they preserve the principal pathway of the electric discharge. Fulgurites are poorly equipped to measure the power of a lightning strike, but are well-equipped to measure the energy of a lightning strike. How do these two differ?

An analogy may be helpful here. Let's say your brother-in-law comes to visit, in between jobs at the circus as a snake-handler. He brings with him his work, which makes your small dog very nervous. He then leaves, having found a new job as a carny operating the toss the ring around the can game. Five weeks later, you fly out to Arizona 1000 miles away to visit your mother-in-law, as you're a sucker for pain. Your brother-in-law is there, too, chilling on the couch and feeding baby rabbits to his python and complaining about the "ordinaries".

In the time since he came to visit, which of these two questions is easier to answer: 1) how far did your brother travel? Or 2) how fast did he go?

Most people would say the first. He traveled at least 1000 miles to go from your house to your mother-in-laws house, snakes in tow. However, you don't really have a good handle on how fast he might have gone. He might have meandered slowly, taking the entire five weeks to travel back to Arizona. Or he may have made it in 8 hours a day of driving over three days.

Perhaps he even flew back to Arizona in a two hour trip. His speed could be anywhere from about one mile per hour (all five weeks) to about 40 miles per hour (three days with 8 hours of driving each day) to even 500 miles per hour (the flight).

Using a fulgurite to constrain the power of lightning is similarly difficult. We can actually work out the energy fairly well from the size of a fulgurite, but have a harder time figuring out the power, since the time over which that energy was put in the ground is less clear. To be certain, it had to happen over the time of the lightning strike or less. We know that had to take about as long as the bolt of lightning, which is to say microseconds to up to a whole second. But that's still a factor of one million in difference.

The energy of lightning is readily recorded in fulgurites. What is this record?

We can measure the amount of energy required to cause a change in a material. For instance, heating a liter of water from its freezing point to the boiling point takes about 100 Calories (about a third of a candy bar), or 418,500 joules. Boiling off this water requires a lot more energy than that, about another 2,260,000 joules. These values are known as the heat capacity (the energy required to change something by a degree in temperature) and the heat of vaporization (the boiling energy). Just like water, rock has well-defined values for both heating events. More specifically, for sand composed of quartz, it requires about 15,700,000 joules for every kilogram of rock vaporized.

How then do we figure out the energy associated with each fulgurite? In this case, we can take the 15,700,000 joules per kilogram value and merely multiply it by the amount or mass of the rock that was vaporized. Luckily, fulgurites are tubes, so the mass of sand that is vaporized can be measured by determining the mass missing when the fulgurite formed. The volume of a tube or a cylinder is equal to the number pi (3.14159) times the radius of the tube squared times the height of the fulgurite. This volume can then be multiplied by the density of the sand, which is the mass of sand per unit volume. This value is typically about 1.7 kilograms per liter of sand, depending on type and physical characteristics of the sand, but usually isn't too far from that value.

Thus, after measuring the volume of the interior of a fulgurite, we can then multiply this volume by the density of sand to get the mass of sand vaporized. Then we can multiply this mass by the vaporization energy to get the total energy preserved (as a minimum) within the fulgurite.

We did exactly this to a set of 266 fulgurites from Polk County, Florida[28]. These fulgurites, which were found at two sand mines, were collected from scanning along the ground for fulgurite fragments. The fulgurites here were formed in white sand, which in this case makes pretty, white fulgurites. All were type I. If we assume that these are typical of type I fulgurites, then the average type I fulgurite records about 50,000 joules of energy, not a lot, relatively speaking. However, the average type I fulgurite is also usually fractured, broken by geologic processes cracking and crunching the soil.

Thus, for nearly all of these fulgurites, we did not actually collect the whole length of a fulgurite. We collected only a fragment, sometimes less than an inch in length. A few whole or nearly whole fulgurites were collected, and these were up to a few feet long. However, with the 266 fulgurites we did collect, we were could calculate a specific energy per unit length. The energy per meter of each fulgurite is then primarily a factor of its diameter or radius.

---

[28] You can find the paper here: https://www.nature.com/articles/srep30586

For now, let's assume this style of measurement is valid. Something interesting comes out from this data.

For most things in nature, we commonly assume a bell curve distribution. A bell curve is familiar to most people who have ever tried to figure out how things relate to averages. A bell curve is a feature of statistics. The field of statistics is kind of tough, and is also a great tool for lying (ranking higher than damned lies). Some parts of statistics are used frequently enough that most of us are familiar with what they mean. The most common statistic used is the average. An average is the value you get when you take some set of measurements, with at total number N, add them all together, then divide them by N. So if you have 10 people, with their weights 120, 134, 156, 157, 176, 185, 193, 207, 221, and 301 pounds, the average is the sum of these (1850 lbs) divided by 10, and the average weight is 185 lbs.

The next statistical value that is commonly used in describing data is the standard deviation. A standard deviation has a bit less obvious of a definition, but, with a bell curve distribution, it is a value that, when added or subtracted from the average value, contains about 68.2% of the data points.

Bell curves pop up often with different data sets, including people's height and weight, as well as intelligence (whatever that happens to mean). It's common enough that the bell curve is called the "normal" distribution.

The normal distribution has such a convincing name that many scientists look for it everywhere. And, being humans, they find it everywhere. This has led to some less-than-correct assumptions about the distributions of things (such as the average power of an earthquake, or wind speed a storm, or height of a flood).

Many physical things don't actually follow a normal distribution, but instead follow a lognormal distribution. Log is short for logarithm, which is something that non-scientists and non-engineers have usually forgotten as soon as they left school. However, they appear commonly in many distributions of physical data, where they often go unrecognized.

Fulgurite energy measurements are one such distribution. If we take the logarithm of the measured energy of each fulgurite (which is just a process that turns the number measured into another number), either as joules or joules per meter, and arrange those values in a distribution, we actually arrive at a bell curve. The consequence of this is that instead of an average (also known as a mean), we arrive at something called a geometric mean. So the "average" energy of a lightning strike was measured to be 50,000 joules, or about 1,400,000 joules per meter (meaning the average length was 3.5 cm). The standard deviation is a bit odd, but instead of a plus/minus value, it is actually a times/divided by value. So the average energy is about 1,400,000 joules per meter, with a standard deviation of a factor of 3. So the average energy recorded in a fulgurite ~68.2% of the time ranges between about 500,000 and 4,000,000 joules per meter.

Lightning burns a lot of energy into the rock to make a fulgurite. We've been able to measure the amount of energy lightning puts into the air as it travels from cloud to cloud or from cloud to ground, and it tends to be one one-hundredth or one one-thousandth of that[29]. Although those differences do seem to be large, the difference is quite explainable by differences in density.

---

[29] Two papers detailing this include: http://onlinelibrary.wiley.com/doi/10.1029/RG017i001p00155/full and http://onlinelibrary.wiley.com/doi/10.1029/RG022i004p00363/full

Air is pretty light and has a density of about 0.001 kilograms per liter. Sand has a density of about 1000× this, 1.7 kg per liter. Therefore, the energy dissipated by lightning in air is about the same as that dissipated into sand, when considered on a per mass basis.

We noted that the energy given is provided as joules per meter, which is necessary in part because the fulgurites are fragmented. How valid is such a measurement actually? Can we really assume each fulgurite to be fragmented in such a way as to preserve the fundamental energy of each lightning strike? The answer appears to be yes.

How do fulgurites break? They are long tubes of glass, so they would probably break much like tubes break. Tubes break roughly like sticks break. Think about where a set of sticks might break. They'd probably break in the middle, because they if you balance a twig on its edges, it'd crack close to the center. Fulgurites should be roughly the same. So when fulgurites break, they do so close to the middle of their length.

These half-fragments of fulgurite themselves may break in two, and other fragments may break further, and so on, until the fulgurite completely disintegrates. Or not. Let's take our example of the stick again. If you break in it in half, then break the half in half, and so on, there will come a point where the stick is so short that it's hard to break any more. Just like sticks, there comes a point where a short length of glass stops breaking in half easily. For fulgurites, this happens when they get about 3-5 cm in length, about 2 inches. They can get smaller, but they can also be bigger. Fragmented fulgurites will average out at about this length.

Because they break in a fashion that can be predicted by halving them, albeit roughly, this breakage can be modeled by computer. In doing so, we were able to demonstrate that fulgurite breakage doesn't change a lognormal distribution of energy per meter for fulgurites: it is instead preserved. Cool!

17. Fulgurite Glass Density

One of the side projects of the energy study of lightning from fulgurites was a measurement of the density of several type I fulgurites. We measured the density just like Archimedes did two millennia ago[30]: measuring their mass, then dipping them in water to determine the volume, and dividing the mass over this volume to get the density. The density of type I fulgurites came in as 1.7 to 1.8 kilograms per liter. This includes the big hollow space in the center of these things. So that means that the fulgurites had the same density—roughly—as the glass they were formed in. What does that mean?

What it means is that lightning, when it strikes sand, completely vaporizes a tube of the sand. It does not melt some glass, then press it outward by high pressure, to form a glassy wall. We know this because, if it did, then we would find that the fulgurites instead have a density close to that of fused quartz glass, or about 2.5-2.6 kilograms per liter.

Instead, it acts almost instantaneously to remove a section of the fulgurite. It melts the material immediately adjacent to this void, but the void itself is shot out as gas. Eventually it does cool and form a solid again, but not inside the tube.

---

[30] Eureka! https://en.wikipedia.org/wiki/Eureka_(word)

18. Fulgurites are Shocking

Recall that there is a difference between energy and power. Power is the energy per time, just like speed or velocity is the distance per time. Returning again to the Delorean in *Back to the Future*, can lightning indeed provide "1.21 gigawatts of power"?

We've now established the energy lightning can deposit into sand at about 1,500,000 joules per meter of fulgurite formed. In order to attain a power, we now need to know how fast that energy is transferred. How long then does a lightning strike last?

We can answer this question pretty well. It's about 1-100 microseconds for a single lightning stroke[31]. Sometimes multiple strokes will occur within a single strike, giving a flashing appearance, but other times there will be only a single stroke making up the strike. Thus, if we take our 1.5 megajoules of energy, and divide it by 1-100 microseconds, the power is between 10 gigawatts and 1 terawatt. A gigawatt is a billion watts, and a terawatt is a trillion. The Delorean would have had enough power from the lightning, and then some, when it got struck. Hollywood wasn't too far off!

Let's think about that power for a bit. A gigawatt is a lot of power. A power company typically can provide about a few gigawatts of power to a city, depending on its average need. The average household needs a few kilowatts, depending on time of day and season, so a power company that is able to provide a few gigawatts can provide power to up to one million houses or businesses. Lightning provides enough power to give about one billion houses power, albeit only for a few microseconds. Sadly, that fact right there is why lightning really isn't a usable power source: a few microseconds of power, distributed all over the world, with no easy way of capturing it, effectively ensuring no likely power grid will ever use lightning. However, this does explain why power grids fail when struck by lightning- too much power will melt and blow protections in the line.

Still, that power is pretty impressive. Gigawatts and terawatts. That power makes a fulgurite. What can that actually do to the ground when it makes a fulgurite?

It turns out that there are a few other geologic phenomena that also have this level of power. A meteor impact is one. Meteor impacts, like fulgurites, drop a lot of power per volume of rock affected. In fact, the scale of power is close between the two. Should we then expect a similar effect between impacts and fulgurites?

Meteor impacts are important in geology. Finding an impact crater can be kind of a big deal. The big one in Mexico[32] has been identified as the one that nixed the dinosaurs. Since then, geologists have often gone hunting for craters, trying to associate them with other extinction events. Extinctions have happened frequently in the past, and several were even bigger than the dinosaur-killer. It would be nice (if a bit scary) if each extinction could be associated with an impact.

So, when geologists have gone looking for impacts associated with extinction events, some geologists have actually found them. In this case, the geologists are looking for impact craters, or the remnants of them. However, sometimes it gets a bit ridiculous, as not every extinction event need be associated with an impact crater, but given the fame of Chicxulub as defeater of the dinosaurs, fame-hungry scientists have gone looking for craters, and finding them, or at least claiming to have done so.

---

[31] See here: http://www.public.asu.edu/~gbadams/lightning/lightning.html
[32] https://en.wikipedia.org/wiki/Chicxulub_crater

Geologists who research impact craters have drawn up 4-6 lines of evidence that can be used to identify an impact event. Geologists who don't research impact craters sometimes will abuse these, or even get them outright wrong, but in such case, the evidence for an impact will typically remain "fringe". If one, or more, of these evidences are present at a site, then it's decently convincing that there was an impact.

The first (#1), and most obvious evidence for an impact is a crater. The first crater recognized to be an impact crater is Meteor Crater in Arizona. This crater, which occurs in sandstone and other sedimentary rocks, also had chunks of meteorite associated with it. Meteorite chunks make up evidence (#2) for an impact crater. If you can find pieces of space rock associated with your hole in the ground, there's a good chance you have an impact!

Not every hole in the ground is in fact caused by an impact. If you've ever been to the big island of Hawaii, and made it to Volcano National Park, you've seen craters that are not formed by impact. In fact, holes in the ground found in volcanic rock are often the most ambiguous when it comes to identifying their origin. However, if it's a hole in a rock that's not volcanic, that may be a good indicator of an impact origin.

The next strong indicator (#3) of an impact is akin to finding meteorites: the detection of high concentrations of platinum group elements. Platinum group elements are a group of 6-ish elements, consisting of ruthenium, rhodium, palladium, osmium, iridium, and, well, platinum. They are all metals that you'd probably like to have a few good solid bricks of, in a vault somewhere if you could. They are, as you might imagine given the name, quite valuable. Platinum group elements, often written short-hand as PGEs, are so valuable because these elements aren't really found regularly on the crust of the earth. When the earth separated into a crust, mantle, and core, these elements sank fast and sank hard. They are in the core of the earth, and we can't really get to them, as drilling through 2000 miles of rock isn't really feasible outside of bad Hollywood movies[33]. So finding any of these elements is indicative of something odd going on.

For asteroids, this does not appear to be the case. Many asteroids, unlike the earth, did not separate into a rocky crust and a metallic core. As a result, they have PGEs present everywhere at roughly the same consistent concentration, including their surfaces. Thus, when an asteroid hits the earth, it drops a rather high mass of PGEs on the earth's surface.

We're not really talking a lot, however. The "iridium layer" as it is known colloquially, is a layer that marks the end of the period of the dinosaurs (called the Cretaceous period, which transitioned into the Tertiary period, giving an acronym for this layer the "KT boundary"). Although the iridium layer sounds cool, the amount of iridium is very small, around about 10 parts per billion. Thus, in order to get an ounce of iridium, you would need to collect over 3000 tons of this soil (or, to get a gram, you would need 100 metric tons). So the "iridium layer" is a misnomer, much like "Killer clowns are everywhere!" that incited local news stations in 2016. Sure, a couple dozen idiots dressed up as clowns to scare people, and that number is more than the zero it should be, but it's not "everywhere".

Detection of a high concentration of PGEs is indicative of an impact. You need at least 10× the background, but not a whole lot more than that. Ten times the background is about ten milligrams per ton of rock. Still, few events really get that high.

The next indicator of an impact is the discovery of tektites (#4). Most of the time these tektites aren't so big. They are better termed microtektites, observable mostly under the

---

[33] https://en.wikipedia.org/wiki/The_Core

microscope. However, glass is rare in the geologic record, and finding it in a geologic layer is indicative of an impact.

The final two impact indicators are both changes that occur to the mineral quartz on impact. The mineral quartz—SiO_2—is one of the most abundant minerals on the face of the earth. Quartz, when subjected to the high pressures and temperatures of an impact, changes on the molecular level. The first change is the formation of something called "planar deformation features". These structures are not visible except under a microscope. They occur when small-scale melting along a specific crystal plane occurs in quartz, forming planar layers of glass. The result is grains of quartz that look like they have hatch-marks scratched through them. These planar deformation features are impact indicator (#5) and are pretty clear-cut indicators of an impact.

The second change is a bit more fundamental to the molecular order of SiO_2. Quartz is the most stable arrangement of Si and O atoms on the surface of the earth, at temperatures typical to the earth's surface (so, from kind of cold to kind of hot, but not volcano-hot). However, when quartz is heated, or squished under high pressure, different arrangements of SiO_2 can arise spontaneously. These are new minerals, and are no longer quartz. They are not really that well-known to non-mineralogists.

If quartz is heated on the surface of the earth, two minerals form from it: tridymite and cristobalite. These minerals both have the formula SiO_2, but are clear to white in color with a different crystal structure. If instead quartz is squished as well as heated, SiO_2 will rearrange into two new minerals, called coesite and stishovite. Both minerals are still white to clear, but are much denser than quartz. In all cases, the continued heating of quartz will form molten SiO_2, independent of pressure.

The two minerals coesite and stishovite are used as the final impact indicator (#6). Since the pressures required to make these minerals are so high, there's not an easy way to make them on the earth's surface except through impact. Thus discovery of these minerals (which really requires a trained mineralogist or mineralogy lab) is a strong indicator of impact. Both point #5 and #6 are indicators of "shock" which is a rapid rise in pressure and temperature due to collision or impact of two or more objects.

So where do fulgurites fall in this? The truth is, we don't know. Even though impacts and lightning strikes both dump a huge—though roughly equivalent—amount of power into rock with equal volume, clear indicators of "shock" associated with fulgurites have not been found. There have been less-than-clear shock indicators found, however. In the formation of planar deformation features, and in the formation of coesite and stishovite by shock, there must be some fundamental changes to the quartz structure. These in-between effects have in fact been observed in fulgurites. As quartz is subjected to shock, there are some permanent, fundamental changes that take place. My work[34], and the work of mineralogist Martin Ende[35], and geochemist Reto Giere[36] (organized in order of publication) all found evidence of something that looked like shock. These were all variations to the structure of quartz at some level.

By definition this evidence is not nearly as convincing as the evidence of shock from planar deformation features or transformation to stishovite/coesite. However, it does provide a

---

[34] With collaborators: http://rsta.royalsocietypublishing.org/content/368/1922/3087.short and https://link.springer.com/article/10.1007/s00216-010-3593-z
[35] http://eurjmin.geoscienceworld.org/content/24/3/499.short
[36] https://www.degruyter.com/view/j/ammin.2015.100.issue-7/am-2015-5218/am-2015-5218.xml

route to forming evidence that is OK at times. Given that some geologists love to find impacts everywhere, sometimes some of them rely on evidence akin to the mid-to-weak evidence we find in fulgurites. Perhaps some of these geologists aren't finding impacts after all, but instead are finding fulgurites formed from lightning thousands of years ago.

19. How old can fulgurites be?

The ability to age-date rocks is one of the triumphs of geology. By determining the exact age of a rock we learn a clear date, an exact note on the history of a rock. Generally these dates are very accurate to two significant figures (in other words, 35 million years, of 440 million years), and can be accurate to three or even up to six significant figures with fancier tools. Determining the exact age of a rock can be done using a variety of methods, though the most successful tend to be based in radioactivity, a process called radiometric dating.

Radiometric dating is a measurement that solves for age by using the composition of a specific set of elements within a rock, arriving at a ratio that gives an age. It is predicated on the idea that rates of radioactive decay have not changed within the timescale of interest. Generally, this appears to be true: outside the cores of stars, unstable elements decay at consistent rates. Even within the cores of stars, they tend to decay at rates that don't change much from constancy.

There are a number of elements used in radiometric dating. The most familiar to the uninitiated is radiocarbon. Radiocarbon decay is not commonly used to date rocks for two reasons. For one, few rocks contain the organic carbon that can be measured using this tool. Secondly, radiocarbon works only to date objects that are 50,000 to 70,000 years old or less, depending on the instrument sensitivity. This places radiocarbon dating firmly within archeology[37]. Indeed, one of the first uses of radiocarbon dating was to show that radiocarbon dates matched the age of a Egyptian boat (the funerary boat of Senwosret III), which had a clearly marked date on its side. Low and behold, the date from radiocarbon matched this date, proving the method.

The dating of rocks is done with methods that invoke the larger part of the periodic table: potassium-argon, argon-argon, rubidium-strontium, uranium-lead, thorium, samarium-neodymium, rhenium-osmium, and a few others. Each of these are used for different rock types, and for different timescales, with some stretching billions of years, and others stuck at tens of millions. Age dating of rocks is contingent on something called a closure temperature, which is the temperature where the elements used for radioactive dating no longer move in and out of a rock. At the closure temperature, elements are frozen in place, and any radioactive decay that may occur produces new elements that are also frozen in place. Buildup of these decay products allows radioactive dating to be done a few hundred thousand to millions of years later.

---

[37] https://www.acs.org/content/acs/en/education/whatischemistry/landmarks/radiocarbon-dating.html

Rock dating

Fulgurites are rocks, and hence it would seem that they should be datable by radiometric methods. Unfortunately, this is not nearly as easy as it might seem. Since fulgurites are heated rapidly, and then cool quite quickly, some of the elements may not have migrated out, effectively making the closure temperature useless, hence the age of the lightning strike may not be determined using radiometric dating. Additionally, since type I fulgurites are relatively pure compositional SiO_2, there are few radiometric methods that measure the ages of these types of rocks because there are few radioactive elements associated with silicon.

Radiometric dating is called an "absolute" dating method. It gives a specific date for the formation of a rock based on the elements found inside of it. This date is something like 44.6 million years old. In contrast, other methods of dating are called "relative". These are based on things that include the positions of rocks relative to each other, or on the fossils found inside of them. Relative dates are less exact and range from "older than that rock" to "40-50 million years old".

One way of dating a fulgurite using relative dating techniques is by saying that the fulgurite had to form since the laying down of the sediment. This timescale may not be well constrained, but typically soils are young (100 to 10,000 years old). Alternatively, some fulgurites are known by the deposits in which they are formed. For instance, we found several fulgurites in Polk County, Florida, in a rock body called the Cypresshead Formation. A "Formation" in geology is a distinctive rock layer from a set time period that is typically composed of the same rock type, generally a sedimentary rock. The Cypresshead Formation is made of quartz sand that stretches from Georgia into Florida and is believed to be a few millions of years old, and was formed primarily as part of river delta. The fulgurites we found in this formation thus have to younger than the age of this geologic formation, since they occurred after the formation was laid down.

In general, an age of zero to five million years old isn't that satisfactory. Hence, amongst the few times scientists have attempted to understand fulgurites, they have used a technique called *thermolumiscence*.

Thermolumiscence dating operates on the idea that crystals, while amazingly regular, are never perfect. Imperfections in the crystal structures of minerals such as sand grains sometimes will trap free electrons that come from any number of sources: radioactive elements nearby, or cosmic rays bombarding the surface of the earth. These free electrons sit in the crystal structure of the altered minerals. When a mineral is heated, these electrons escape, giving off a slight glow. The intensity of this glow is proportional to how long a mineral has been sitting, exposed to this radiation. On the earth's surface, this timing is related to how long a mineral has been exposed to sunlight.

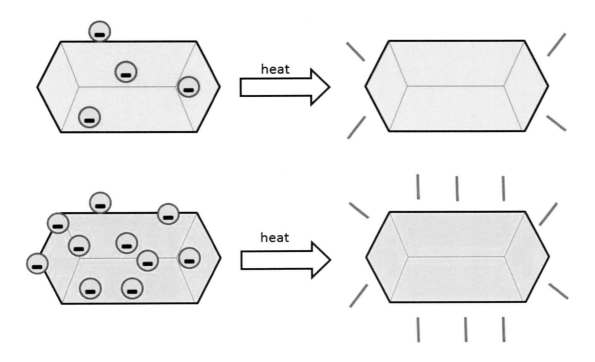

Thermolumiscence is based on the glow of a mineral: more glow means more electrons, meaning longer age of exposure to sunlight.

Each mineral can be "reset" by a heating event. If heated to hot enough of a temperature, the electrons bounce out of these mineral imperfections. Thus, a heating event, such as a lightning strike, resets this counter. A fulgurite can be dated thus by the amount of electrons it loses, or glow it emits, when heated in the laboratory. Rafael Navarro-Gonzalez used this technique to identify the age of a fulgurite from the Sahara. Then, using gas chemistry trapped within the fulgurite bubbles, Rafael and his colleagues identified the species of plant present when the fulgurite was formed. Intriguingly, these results, which came from a fulgurite estimated as 15,000 years old, suggested that the Sahara desert was much richer in vegetation than the present. In other words, the Sahara, as a desert, has gotten much bigger and hotter over time, resulting in less vegetation.

Another route to estimating the age of a fulgurite is by its "glassiness". Glass is an unstable material, and slowly converts from its disordered state to a more ordered mineral state. The rate at which a glass devitrifies—or loses its glassy nature—can be determined by chemical rate laws. A chemical rate is the speed at a chemical reaction occurs. Some can be very fast (less than seconds), and others much, much slower (billions of years). The rate of devitrification depends on what a glass is made of. Of the four types of fulgurites, the most stable are the type I fulgurites, formed in quartz sand. SiO_2 is slow to devitrify, taking millions of years under some conditions. In contrast, glass that contains other elements in addition to Si and O, such as sodium, potassium, calcium, or magnesium will devitrify relatively quickly. These glasses transform into clay minerals upon the addition of water. The exact rate of these devitrification reactions is not known, but likely varies from tens of years to tens of thousands, depending on the composition of the fulgurite. It is unlikely that a fulgurite will be stable for millions of years if it is less than 90% SiO_2.

As a result, the age of a fulgurite is directly dependent on its composition. A type I fulgurite will be between zero and a few million years old, whereas a type II fulgurite will be between zero and a few thousand years old. Type III fulgurites are less easy to constrain, but may be stable for long periods of time, like the type Is. There is some alteration of fulgurites as they age, for instance the type I fulgurites become much more wispy and fragile as they slowly devitrify.

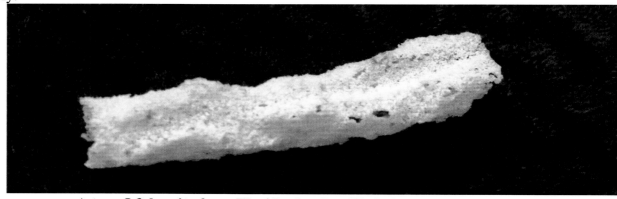

A type I fulgurite from Florida that has likely begun to devitrify

Some fulgurites have been found in old rocks, and fulgurites in rocks tend to be much more stable than other fulgurites in the long run, since rocks themselves tend to be quite stable. As an example, some fulgurites from the Adirondacks in New York state were formed many years ago, and persist even today.

A fulgurite from Ampersand Mountain, near Harrietstown New York. Picture taken by my colleague Mark Stewart.

Thus the age of a fulgurite is hard to estimate well. The one technique that works well—thermolumiscence—is kind of expensive, and may not work well for all fulgurites. A fulgurite's age may be estimated from its composition: a type II fulgurite is probably 100 years or younger, whereas a type I and IV fulgurite may be thousands to millions of years old.

20. What are some of the minerals in fulgurites?

We've already discussed some of the more unusual minerals that occur in fulgurites, such as iron metal, iron silicides, and metal phosphides. Beyond those, most fulgurites are composed of glass, which generally isn't considered a mineral.

The most common minerals that are found in fulgurites are the SiO_2 minerals. These minerals include the one that pops up on clocks: quartz. We've discussed this mineral a few times earlier, and that's because quartz is a mineral that is found almost everywhere on the surface of the earth. We've mentioned a few other variations of SiO_2 in passing as well: cristobalite, tridymite, coesite and stishovite. These are all compositionally identical to quartz, but vary in crystal structure. Each of these minerals is indicative of specific pressure-temperature environment.

Mineralogists—scientists who study minerals—use a tool called *phase diagrams* to understand what minerals tell us about geologic conditions. A phase diagram is a relationship between two different changeable things, termed variables, and the mineralogy expected to occur under those variables. One of the most common phase diagrams is a pressure-temperature phase diagram. The phase diagram for SiO_2 is shown below.

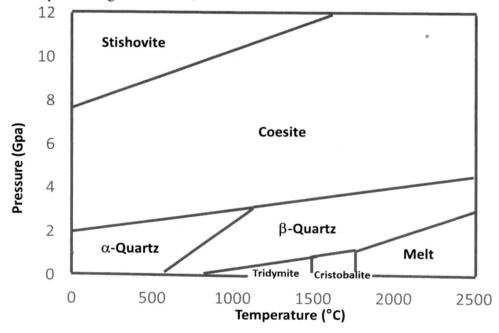

The phase diagram of SiO_2 shows [38] how specific minerals relate to pressure and temperature. The temperature is given in Celsius, which is the temperature scale used pretty much everywhere outside of the USA, and the pressure is given in gigapascals (GPa). A pascal is a relatively small unit of pressure. Your body is subjected to about 100,000 pascals just because you're under the earth's atmosphere right now. A gigapascal is one billion pascals, or 1,000,000,000 Pa. Thus, dividing a gigapascal by an atmosphere's worth of pascals means each GPa is 10,000 atmospheres of pressure. That's not really useful. Probably a bit more useful is

---

[38] http://www.quartzpage.de/gen_mod.html

the fact that a GPa is the pressure felt by a rock that is buried under 30 km of rock (20 miles). That's a lot and would certainly flatten a person.

Quartz transforms at high temperature and low pressure to the mineral tridymite, and then at somewhat higher temperature tridymite turns into cristobalite. Neither mineral is really that familiar to a casual mineral collector, though "snowflake obsidian" has cristobalite[39]. At high pressure and middling temperature, the minerals are coesite and stishovite, which we previously discussed in impact rocks. At high enough a temperature, all of these minerals melt, and when cooled can form glass, termed lechatelierite.

Fulgurites are erroneously termed a variety of lechatelierite. Although many fulgurites contain lechatelierite, a fulgurite is a rock, whereas lechatelierite is more akin to a mineral (though not really- it is a glass, after all). The type I fulgurites are composed mostly of lechatelierite, hence some websites will lump all fulgurites in with lechatelierite.

Quartz tends to be the most common mineral in many fulgurites. In fact, only a few fulgurites have ever been reported that have no quartz or lechatelierite. Most of these occur in rocks as type IV fulgurites and the rocks with these fulgurites themselves have no quartz to begin with.

Most other fulgurite minerals are less common than the SiO_2 minerals, as many melt at low temperature. With a low melting point, minerals that were present in the original target rock transform into glass and may mix or flow, losing specifics of their composition. Some minerals may be stable at higher temperature, but tend not to be as common. One that we have seen in fulgurites is the mineral zircon. Zircon is the mineral that is the major carrier of the element zirconium, which is otherwise not well known. The mineral has a chemical formula of $ZrSiO_4$. This mineral melts at a pretty high temperature. Lightning is hot enough to melt it, and we saw melted zircons in some type I fulgurites.

However, zircon is also stable enough to not necessarily melt under some conditions. Instead, zircon breaks down to form a new mineral, called baddeleyite. This mineral has a formula ZrO_2 (and when recrystallized in the lab, gives the cheap diamond substitute *cubic zirconia*). Baddeleyite is formed by a process called *incongruent melting*. In incongruent melting, a mineral melts to give a new mineral, and a liquid. In some fulgurites, zircons melted by this process, turning into new minerals. In contrast, *congruent melting* occurs when a mineral completely transforms into a liquid with the same composition as the starting mineral. Fulgurites can have examples of both, and the melting fate of minerals is dependent on where in the fulgurite they are found. Congruent melting occurs in the interior of a fulgurite, whereas incongruent melting is more common to the exterior.

---

[39] http://www.charmsoflight.com/snowflake-obsidian-healing-properties.html , with metaphysical properties listed

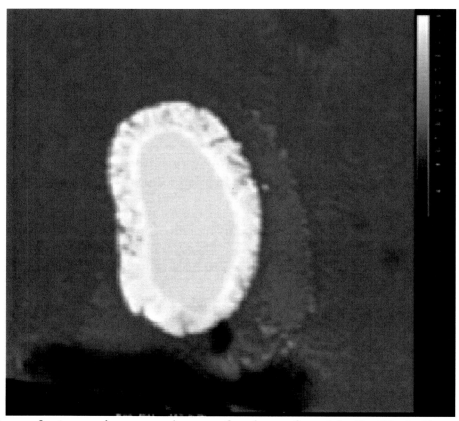

This is an electron microscope image of a zircon found in the York County, PA fulgurite. The interior of this mineral is made of zircon, and the bright, cracked exterior is baddeleyite. It is surrounded by glass. The width of this mineral grain is 10 μm (about a fifth the width of a human hair).

Two other minerals occasionally found within fulgurites are graphite and mica. Both of these are highlighted because they suggest interesting geochemical things are occurring when they're found. Graphite is well-known to most of us as being the precursor to diamond, and is composed of the element carbon. It is found in a few fulgurites and indicates that some organic matter, such as a tree root, ignited when the fulgurite was formed. To date no diamonds have been found in fulgurites, outside of a bit of diamond grit that once got caught up in a fulgurite we cut for analysis.

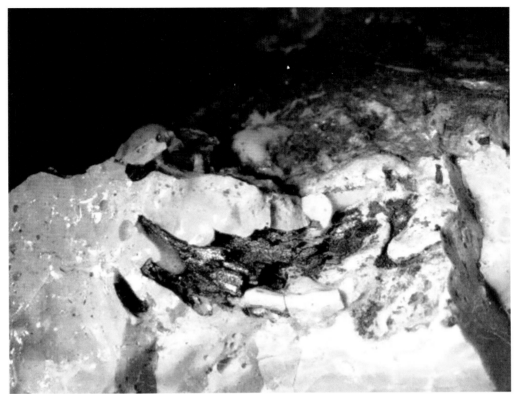
Graphite in the Wisconsin Blue fulgurite. It kind of looks like charcoal, because it forms from charcoal.

The mica minerals are fun. The mineral muscovite forms cool thin sheets that separate quite well when pulled apart with a fingernail. It is translucent or even transparent when suitably thin, and has historically been used as window material to check on flames in furnaces as it also is a good thermal insulator. Mica, which includes a number of other minerals including muscovite, is a mineral called a sheet silicate. These sheets give mica its unique character. Mica also is a mineral that contains water as a constituent. Water has a much lower boiling point than most minerals, and when mica is heated, this water can boil away and melt the mineral. In fulgurites, some micas show distinct evidence of water boiling out of their chemical structures.

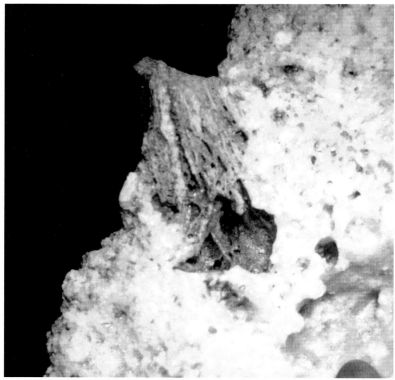

Micaceous rock fragment in the fulgurite from Portugal. Note that the upper part of this rock is layered, but bubbles and splits the closer it gets to the fulgurite glass.

There are a number of other minerals that occur within fulgurites, but these are some of the more unusual ones. Fulgurites do present an opportunity to understand how minerals melt, especially when time is a factor in their melting. Most minerals melt congruently in fulgurites, because the melting timescale is so fast.

21. What type of rock is a fulgurite?

There are three main rock types. Most of us remember that fact, as most all people with an education have taken some geology at some point, and the three rock types are first taught in elementary school. These three rock types are sedimentary, igneous and metamorphic. Reviewing our quick definition before, igneous rocks come from molten rock, sedimentary rocks are particles of rock that are glued together, and originate from sediments or from minerals precipitated from the sea, and metamorphic rocks are formed from heating and/or putting other rocks under pressure making new minerals, textures, and crystals. We previously stated that fulgurites are most commonly grouped in the metamorphic rocks, as they are rocks transformed by the high temperature and possibly high pressure environment initiated by lightning.

Fulgurites are best grouped with a type of rock called "Pyrometamorphic" rocks. This name, which may sound like it comes from a comic book villain, means metamorphism due to fire. In general, pyrometamorphic rocks occur only in a specific environment: when lava flows through a carbon- or organic-bearing (e.g., coal-bearing) rock. As the lava flows through such a rock, it spontaneously ignites the carbon, which heats it rapidly and changes the mineralogy significantly. Some very unusual rocks result, and are named "buchite", "porcellanite", and "emery". The latter two may invoke specific imagery. Porcellanite is a white to yellow rock with a texture to that of porcelain, and is composed in part of similar material. Emery is made of hard minerals, including corundum, that make up emery boards. Buchite is a natural glass we have mentioned before. More details on pyrometamorphic rocks can be found in Rodney Grapes' book, "Pyrometamorphism".[40]

Fulgurites differ from most pyrometamorphic rocks in that they are formed quite quickly, and the source of heat is not burning carbon, but lightning. Indeed, by definition fulgurites are quite problematic to place within a rock type using the easy definitions. Calling fulgurites metamorphic is not problematic, but fulgurites also form from molten rock, which places them in the igneous category of rock. This could be superseded by requiring the molten rock to either be subsurface or expelled from a volcano, but under the broader definition, a fulgurite is also an igneous rock. Finally, fulgurites entrain grains of relatively unaltered sand within their glass, and hence, by the definition of "gluing particles of rock together" fulgurites could also be called sedimentary rocks. This might be avoided by requiring the gluing process to proceed by water and dissolution, but under the broad definition, a fulgurite is a sedimentary rock as well.

These quirks are also true of pyrometamorphic rock (albeit less so for "sedimentary"), and of impact-formed rock such as tektites. In the case of both pyrometamorphic and impact-formed rocks, both types of rock are found with other rocks that obviously did not reach the melting point of the rock- these are volumetrically the majority of both pyrometamorphic rocks and of shock-metamorphosed rocks. Comparatively, molten rock is generally most of the mass of fulgurites. Fulgurites thus fall in a bit of ambiguous "other" zone for rock type. Pleasantly, this is a bit of trivia that may stump the geology professors at your local university!

---

[40] You can find said book here: http://www.springer.com/us/book/9783540294542

22. What is the difference between natural and artificial fulgurites? When might that be important?

A natural fulgurite is a fulgurite formed by lightning. When people talk about fulgurites, they generally mean these types of fulgurites. Most of the time natural fulgurites occur in sand or soil. However, sometimes lightning may strike a material that's kind of odd, such as a sidewalk or black-top, forming very unusual materials. Such fulgurites we have termed "Anthropogenic", indicating they were formed in human-influenced material. Some examples of anthropogenic fulgurites include fulgurites that have melted copper wire in them, or are made partially of sidewalk.

This fulgurite, from near Dallas, TX, was formed when a downed power line sat and melted on the embankment next to a road. Note the large copper sphere from the power line.

In general, geologists care only about rocks that are found in nature. Artificial rocks, like those made in the lab, are useful primarily as analogs. Unintentionally formed rocks in manmade substances are rarely encountered, and generally considered to be scientifically worthless. Anthropogenic fulgurites are thus almost never considered in the geologic literature, with some justification as the mineralogy that forms in some of these things is mostly dependent on the addition of non-natural material.

A separate type of fulgurite we have termed "Artificial" fulgurites. These fulgurites form when a man-made electric discharge, for instance from a downed powerline, melts and vaporizes the rock with which it comes in contact. An artificial fulgurite may occur in natural material, which can make distinguishing it from a natural fulgurite tricky, especially after the power line culprit has been removed.

Here's a scenario where distinguishing between a natural and an artificial fulgurite might be important. Suppose there is a high voltage power line in the woods. It is drought season; it has been dry for some time, and the trees are surrounded by dry brush. The power line, with its wooden pole poorly tended from the late 1940s, finally succumbs to termites, and cracks suddenly. The power line snaps and the line carrying high voltage electricity drops and ignites some of the brush, and then finally settles onto the rock that anchored the power line. The brush fire turns into a forest fire and destroys houses and property. The owners of the property sue the power company for negligence. In discovery, firefighters and engineers find a fulgurite where the power line might have come in contact with the ground. Can the plaintiffs prove the power company was to blame?

This question is way trickier than it may seem. Fulgurites are found naturally almost anywhere. Differentiating between a natural and an artificial fulgurite hence relies on some of the subtleties of fulgurite formation.

First, most artificial fulgurites are also anthropogenic fulgurites, at least to some degree. Power lines are made either of copper or aluminum metal, which are great and good conductors of electricity, respectively, and which are also expensive and cheap, respectively. Copper metal melts at about 1085 °C. Aluminum metal melts at 660 °C. The temperature required to melt rock ranges from 1150 to 1700 °C, with the latter being applicable to quartz-rich rock. Thus, if a power line is in contact with the soil, delivering current, it will be in contact with rock that is much hotter than the melting point of the power line. It **will** melt.

The presence of melted conductor wire is hence an important line of evidence for an artificial fulgurite. Two other lines of evidence may help distinguish natural and artificial fulgurites, and are based on key difference between how these two types of fulgurites form.

As we've stated before, lightning is fast. There's probably a proverb or simile that says as much. Since lightning is so fast, it drops a lot of energy in a rock over a very short time. Compare that to a power line. A power line has a lot less power than lightning, as lightning is more powerful than every type of sustained human power source available. In order to make a fulgurite, a power line must be in sustained contact with the ground—soil or rock—for some amount of time, ranging from minutes to hours. This sustained contact generates differences between artificial fulgurites and natural ones.

A long duration for an electric arc means that the rock is constantly being heated for a long period of time. Lightning, heating over microseconds to milliseconds, does its work and goes away, vaporizing a lot of rock, melting some, and baking only a little. Sustained heating by a power line means that heat has longer time to travel away from the arc, melting more of it, baking some of it and vaporizing a little of it. The consequence of this is that natural fulgurites are small, have large void spaces (where the rock vaporized) and generally lots of glass. Artificial fulgurites have a lot of baked material (not quite melted, but dried out and clumped together), and a lot less void material. The shape helps distinguish these two types of fulgurites.

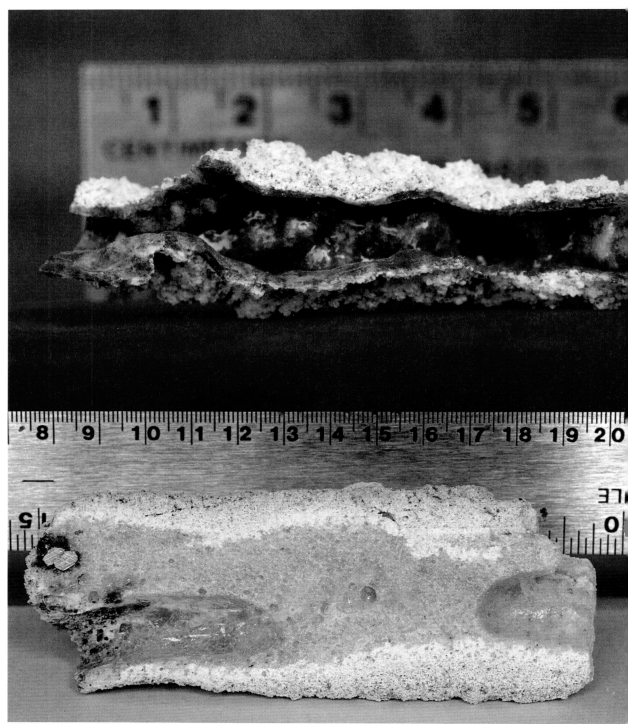

As an example, these are two fulgurites that formed in quartz sand. The top one was formed by natural lightning in Florida. On the bottom is one was formed in quartz sand by a downed power line (on the left is a small fragment of the aluminum conductor). Note the differences in size, in glass shape, and in the amount of hollow void space.

The third distinguishing factor has to do with the time of contact. An artificial fulgurite is heated for a long period of time compared to lightning. This long duration of contact means that some chemical reactions that take a while to go are more likely to go in power line fulgurites vs. lightning fulgurites. An example of this transition is the formation of cristobalite from the mineral quartz. Both minerals have the formula SiO_2, as we have discussed before. Cristobalite packs its atoms in a different way than quartz, making a less dense mineral than quartz. This conversion takes time, requiring at least a few minutes at high temperature, and hours at lower temperature. A natural fulgurite, formed over seconds, will rarely have cristobalite as it is heated and cools too quick. In contrast, an artificial fulgurite should have cristobalite. These two minerals are not easy to distinguish by sight, but can be identified relatively easy in a mineralogy lab using non-destructive techniques.

Fulgurite formation pathways may be part of forensic science for some important legal cases. The differences between artificial and natural fulgurites arise as a consequence of the heating timescales of the fulgurites.

23. What future science might there be for fulgurites?

As you can see, there's a lot of unusual things that happen with fulgurites from strange new minerals to phosphorus chemistry alteration to detailing the physics of lightning to challenging our definitions of rock types. Fulgurites are one of the few varieties of glass that occur in nature, hence they have potential for adding new data to understanding rocks, geochemistry, and the environment.

Much of the future science of fulgurites will likely circulate around individual rock analysis. This can be summarized as "Hey- look! There's a cool rock over here! Let's study it and tell people what we find!" In general such findings improve science incrementally- they may detail new chemical reactions or mineral changes, but generally don't change our understanding of rocks. However, Essene and Fisher did this with their Winans Lake fulgurite, and that resulted in a major scientific breakthrough with its discovery of extremely reduced minerals within the fulgurite. Thus these sorts of findings are important to the body of scientific literature.

Some of the new areas of research that may benefit from fulgurite research are those areas that could use time constraints. Fulgurites hold much potential in the area of time- and temperature-dependent elemental and mineral analyses. Most geologic systems are heated and react over the course of thousands and millions of years. Fulgurites, in contrast, are formed on the order of seconds to minutes. Thus mineral transformation timescales can be studied in depth with fulgurites. If a reaction takes minutes to form a new mineral, some fulgurites may have the new mineral and others may not. Additionally, elements may redistribute themselves, and others may not over the fulgurite formation timescale. Our understanding of these reactions may benefit strongly from fulgurite analysis.

An important part of future fulgurite research will be better clarity on the pressure that is associated with lightning. As we discussed before, there is overlap between impact produced glasses and fulgurites. It is important to understand how pressure accompanies a lightning strike to know when fulgurites may produce similar structures as those found within impacts. We know lightning is high pressure because of thunder (thunder is a pressure wave associated with turning air into plasma), but how much this pressure wave goes through soil is unclear.

The science of geomagnetism may benefit significantly from fulgurite studies. Lightning induces magnetic fields within the rock it strikes, so much so that a compass could be used to identify an ancient lightning strike. Additionally, lightning strikes have been discussed as one of the main routes to forming natural magnets (and hence allowing all of early sea navigation to take place!).[41] Understanding how lightning affects rocks to change their magnetic field, in addition to understanding the magnetic fields of fulgurites will both elucidate the geologic and magnetic history of the earth's surface.

Fulgurites also present a new way of measuring some of the characteristics of lightning. In one of our previous works, we demonstrated how fulgurites capture the energy of lightning, and what they tell us about the distribution of lightning energy. When this is coupled to more detailed measurements of lightning, it may be possible to determine several other lightning features, perhaps including some time constraints. This would be done with inclusion of element distribution and mineral formation timescale measurements.

Probably the most unusual area in future fulgurite research will be the search for quasicrystals in fulgurites. Quasicrystals are an obscure material that were discovered recently

---

[41] See lodestone, the first, natural magnet: https://en.wikipedia.org/wiki/Lodestone

(in the 1980s). A quasicrystal is similar to a crystal, but differs in some key respects. Recall that minerals are crystalline. Crystals are a repeated, ordered structure of atoms. Can something be ordered but not repeating? It turns out the answer is Yes! Such structures are called "quasicrystals" since they are almost crystalline (ordered) but not repeating. The first quasicrystals were made by depositing metals from vapor on surfaces, and had structures akin to something called the Pembrose tiling, an ordered but not repeating way of arranging parallelograms around 5 and 10-point structures.

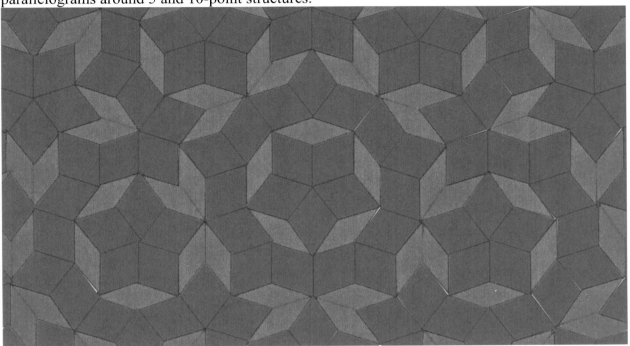

You can make your own Pembrose tiling by taking two types of parallelograms- one with an angle of 72° and 108°, and the second with an angle of 36° and 144°, and arranging them as you will. You may end up with either a regular or irregular tiling pattern. You can compare these two the atomic structure of quasicrystals.[42]

The first quasicrystals found in nature were discovered, appropriately, in meteorites.[43] These quasicrystals, which were composed of iron, nickel and aluminum metal, were not abundant. Indeed, they were present as small micrometer grains within other material in the meteorite. In order to form, the minerals had to form by quickly and cool quickly, and were attributed to impact melting of the meteoroid (from which the meteorite came in space). Recall again that we said fulgurites are similar to impactites. Indeed, Abby Sheffer reported an aluminum-iron-silicon phase with a fulgurite from Massachusetts, and we also reported iron-silicon phases within several of our fulgurites. Thus fulgurites may prove to be one of the first natural, terrestrial sources of quasicrystals.

---

[42] https://en.wikipedia.org/wiki/Quasicrystal
[43] Discussion of the paper is here: https://phys.org/news/2016-12-khatyrka-meteorite-quasicrystal.html. The paper itself is here: Bindi, L., Yao, N., Lin, C., Hollister, L. S., Andronicos, C. L., Distler, V. V., ... & Steinhardt, W. M. (2015). Natural quasicrystal with decagonal symmetry. *Scientific reports*, *5*, 9111.

Predicting scientific advances beyond these research areas is part guesswork and part positive thinking. It's hard to say where science will end up, but given the recent path of scientific discovery, the above do appear to be the most likely new areas of discovery in fulgurite research.

24. Non-scientific uses

By and large, fulgurites have mostly scientific value. They are not useful as ores and they don't really have unique physical properties on the macroscopic scale. Their industrial value is minimal.

There is a larger subculture that uses fulgurites for metaphysical purposes. This is due largely to the "WOW" factor once a person knows what a fulgurite is. Metaphysical properties are by their nature untestable by science. But like many things, interest in metaphysical properties can bring people into science, hence there's value for this from the context of improving scientific literacy.

The primary non-scientific use of fulgurites may eventually be for litigation and insurance claims. As such, a fulgurite may serve as evidence. Fulgurites are generally formed by lightning, but their formation by other electrical sources is clear. The discovery of a fulgurite may help determine legal fault in insurance cases and class-action lawsuits. The forensic evidence for fulgurites likely relies upon distinctions between natural and artificial fulgurites, as outlined in the prior chapters.

25. Individual fulgurite discussions

I currently have a collection of about 50 fulgurites from separate locations. Some have more than one fragment, others are limited to a single individual. In this section I'll show pictures of some of these and will try to provide some of the basic science behind them.

<u>Polk County, FL</u>. By far most of my fulgurites originate from a pair of sand mines in Polk County Florida. We examined these in depth in 2016, demonstrating the preservation of a lognormal trend for lightning energy. Nearly all of these fulgurites are type I fulgurites, and compositionally they are all >98% SiO_2. The sand mines from these locations produce abundant fulgurites, and with a good eye, a collector can find dozens in an hour. Most are small fragments, but occasionally larger individuals can be found. One very large individual, about a foot long and 3 inches wide, was found by a grad student working with me in 2010. We've looked at several of these fulgurites and not one has cristobalite (all quartz), though a few have some other, unusual phases in them. The sand mines probably have several millions of years' worth of fulgurites, which is why they are so abundant at this local topographic high. Florida is not a tall state, but these mines are located on the Lake Wales Ridge, one of the tallest points in the entire state (about 300 feet above sea level). Thus these "Florida Mountains" have received a lot of lightning over the few million years they have been exposed as sand dunes.

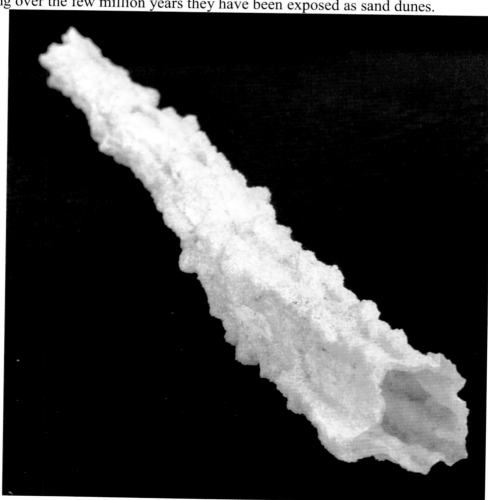

Polk County Fulgurite. Length is about one foot.

Galveston TX. The only other fulgurite I have found in person comes from a sidewalk in Galveston Texas. While at a conference, I wondered off to get some exercise at a break, and happened across this fulgurite. I've not done any detailed analysis of it, but its blue coloration is likely due to mixing of carbonate from the sidewalk into the glass of the fulgurite. I've seen similar light blue coloration in fulgurites from Dallas Texas as well.

Galveston, TX Fulgurite, as found in situ.

Fulgurite and fulgurite path for the Galveston, TX fulgurite (a type II).

York PA. One of the most scientifically interesting fulgurites I have seen comes from York County Pennsylvania. The site, which is on the collector's parent's property, had a small fire break out during a thunderstorm in 2004. It quieted down a few hours later, and the collector went out and found several fulgurites associated with the fire. I purchased these off eBay, and the collector kindly provided me with soil and nearby pictures. When the fulgurite was cut for analysis, we happened upon several distinct metal grains scattered throughout the fulgurite. One exogenic droplet fulgurite was associated with the fulgurite, and when we cut that one in two, we saw a large (~1 cm) metal grain that was formed of all the weird Fe-Si minerals outlined above in the "rule of three" section. This fulgurite provided a small bonanza of unusual minerals in its glass.

York County fulgurite. This fulgurite, a type II, had many side vents.

Greensboro NC. A fulgurite with less distinctive mineralogy came from Greensboro North Carolina. This fulgurite, which was featured on the cover of Nature Geoscience back in 2009[44], was quite colorful with deep blues interspersed with reds and browns. Unfortunately, the collector wasn't the most responsive for details on the fulgurites (purchased off of eBay), and we ended up only with some samples of the nearby soil after contacting a local university soil scientist. This fulgurite, which was composed mostly of iron, aluminum, and silicon oxides, happened to have several unusual quartz grains inside of it. The quartz grains, when analyzed by a lab at the University of Bradford in the UK, suggested that the rock had been shocked when the lightning struck it. We searched for other indicators of shock but were not lucky enough to confirm this finding.

The Greensboro, NC fulgurite is a Type II. Some of the glass was blue, but most was black.

---

[44] http://www.nature.com/ngeo/journal/v2/n8/index.html has a good, colorful picture of this fulgurite

Tucson AZ. One of the more unusual fulgurites I acquired came from a rock shop in southwestern Tucson Arizona. This fulgurite, which was retrieved by a collector and sold to the dealer in 2006, was fairly massive, but heavily fragmented. The fulgurite happened to have some unusual structures on its surface, including a trace of a lightning strike along a pebble that got caught in the fulgurite. Additionally, the fulgurite was obviously anthropogenic- a huge blob of aluminum metal that had been melted when the fulgurite formed was connected to a few pieces of the fulgurite. We have not analyzed this fulgurite in depth, but there do appear to be other metallic minerals also present within this fulgurite.

General picture of the type II Tucson AZ fulgurite.

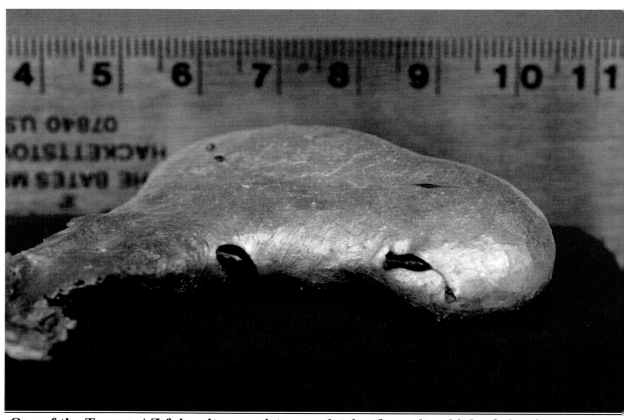
One of the Tucson AZ fulgurites consists completely of a molten blob of aluminum, with small droplets of black glass embedded within.

Sand Hills NE. One fulgurite I acquired off of eBay proved to be most unusual. This fulgurite, which came from the Sand Hills sand dunes of Nebraska, was composed primarily of SiO_2. However, structurally it was vastly different than other type I fulgurites. Why? It turned out that this fulgurite actually formed in part due to a downed power line. There were melted globs of aluminum in one piece, and with a more detailed analysis, we saw the quartz had transformed in part to make the mineral cristobalite. This provided a good test sample for distinguishing between natural and artificial fulgurites.

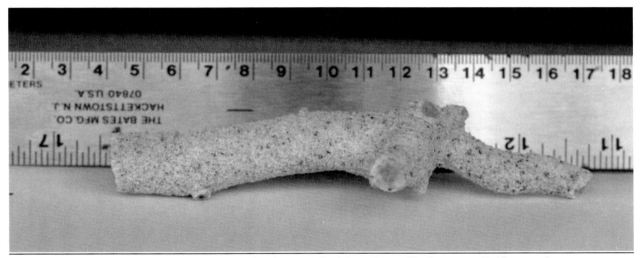

The Sand Hills Nebraska fulgurite is could either be called a type I (due to its composition of nearly pure SiO_2), or a type II, due to its blockiness. This ambiguity is likely due to its formation by a downed power line.

Portugal. I acquired one large batch of fulgurites from the Tucson Gem and Mineral show in 2009. The fulgurite came from Portugal, and I later found out that they were associated with a study done by Martin-Crespo that year.[45] This fulgurite, which was one of the largest ever seen, had some peculiarities to it. For one, it had a lot of cristobalite. Second, when formed, it was said by the finders that they could sense static electricity and heat for days after a thunderstorm had passed through. Additionally, the fulgurite had formed in a power pylon, and steel from the pylon had got stuck in the glass, in addition to the fill used around the pylon. Martin-Crespo's work was the first to analyze an anthropogenic and likely artificial fulgurite! Our own studies of this fulgurite were focused on about ten large samples with varied composition, but Martin-Crespo's original work should be seen as the leader on this fulgurite.

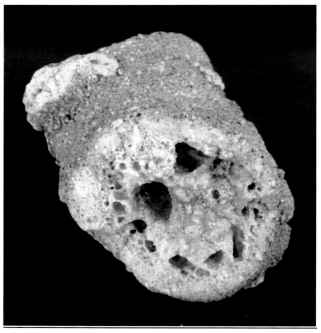

The Portugal fulgurite, a type II, is quite heterogeneous. The white glass has a faint green tint to it, but other sections are dark black.

---

[45] The paper is here: http://eurjmin.geoscienceworld.org/content/21/4/783

<u>Wisconsin</u>. Perhaps the most interesting fulgurite in my collection comes from Wisconsin. This fulgurite was composed of a beautiful blue glass, which ranged in color from sky blue to a deep navy blue. We acquired this fulgurite from ebay, and the collector was kind enough to provide soil samples associated with the fulgurite. We are still searching for the reason for this blue coloration, and hope to publish our findings in the scientific literature shortly.

The Blue Wisconsin fulgurite. Blue coloration ranges from light blue to dark blue (more accurate names can be found in your local paint shop for variations in blue color…)

26. Conclusion

An artist appreciates the aesthetic. Aesthetic is defined in part as "feeling"; for proof, consider its opposite: anesthetic. Scientists have a reputation more for the latter than the former. Consider Walt Whitman, who describes how the sense of beauty is lost when scientists discuss stars using numbers and charts[46]. Or the common complaint of the middle school student, "When will we use this in the real world?"

Our society is blessed and fortunate that we can employ people to discover the nature of nature and the fundamental truth (math?) of the universe. It does so with an expectation: that such study may at some point yield fruit. The discovery of electricity and the description of its properties both occurred well in advance of the first electric lights or power stations, but they were considered worthy of study. The findings of geology protect our homes; new research in chemistry heals us of illness; physics provides an abundance of new knowledge employed in modern communications. But yet, science and art are not considered to overlap.

In Walt Whitman's poem, there is a sense that understanding something robs it of its beauty. Isaac Asimov[47] has a wonderful rebuttal of this: knowledge of "something" enhances the wonder we feel when we consider that "something". For instance, the night sky because more wondrous when we consider other planets orbiting stars, vast bountiful geologic features barely imaginable here on the earth, or the potential for other lifeforms looking back at us. Science has given us this; the mere viewing of the night sky only provides us with a few white dots on a sea of black.

Scientists are not known as artists, yet scientists are all seeking the aesthetic: the sense of joy unveiled when unlocking part of the natural world. As students of science, there is a sense of wonder when we come to a deeper understanding of something. Consider this: if you are a shown a fulgurite, without knowledge of what they are, you would probably be unimpressed. It's just a tube-like thing that's kind of brown. Then, when you learn they were formed by lightning, they become more interesting. Hmm- that's cool! Now I can associate these with those awesome storms that strike in the summer. When you continue to grow in knowledge about these rocks, eventually you'll start thinking how similar they are to space rocks, or how a particularly thick fulgurite may in fact be showing you evidence for a powerful lightning explosion that would have been heard from miles away.

Fulgurites are interesting because of what we know them to be. A rock that captures the awesome power of lightning, frozen in time and disordered in glass, certainly merits our understanding and our care as enthusiasts of the natural world.

---

[46] https://www.poetryfoundation.org/poems-and-poets/poems/detail/45479
[47] https://www.washingtonpost.com/archive/entertainment/books/1979/08/12/science-and-the-sense-of-wonder/679c0f9c-6690-45c1-b3f4-7172463a5f76/?utm_term=.4216cd8dbdd2